Participative Web and User-created Content

WEB 2.0, WIKIS AND SOCIAL NETWORKING

OECD

ORGANISATION FOR ECONOMIC CO-OPERATION AND DEVELOPMENT

The OECD is a unique forum where the governments of 30 democracies work together to address the economic, social and environmental challenges of globalisation. The OECD is also at the forefront of efforts to understand and to help governments respond to new developments and concerns, such as corporate governance, the information economy and the challenges of an ageing population. The Organisation provides a setting where governments can compare policy experiences, seek answers to common problems, identify good practice and work to co-ordinate domestic and international policies.

The OECD member countries are: Australia, Austria, Belgium, Canada, the Czech Republic, Denmark, Finland, France, Germany, Greece, Hungary, Iceland, Ireland, Italy, Japan, Korea, Luxembourg, Mexico, the Netherlands, New Zealand, Norway, Poland, Portugal, the Slovak Republic, Spain, Sweden, Switzerland, Turkey, the United Kingdom and the United States. The Commission of the European Communities takes part in the work of the OECD.

OECD Publishing disseminates widely the results of the Organisation's statistics gathering and research on economic, social and environmental issues, as well as the conventions, guidelines and standards agreed by its members.

This work is published on the responsibility of the Secretary-General of the OECD. The opinions expressed and arguments employed herein do not necessarily reflect the official views of the Organisation or of the governments of its member countries.

Corrigenda to OECD publications may be found on line at: *www.oecd.org/publishing/corrigenda*.

© OECD 2007

Foreword

This report is one of a series on digital broadband content prepared since 2005, focusing on changing value chains and developing business models and the implications for policy. The series is part of ongoing OECD analysis of the digital economy and information and communications policy. The report was drafted by Sacha Wunsch-Vincent and Graham Vickery of the OECD Directorate for Science, Technology and Industry as part of the digital content series under the overall direction of Graham Vickery (Head, Information Economy Group).

The report was presented to the OECD Working Party on the Information Economy (WPIE) in December 2006, and in March 2007 the Committee for Information, Computer and Communications Policy recommended that the report be made available to the general public. The authors are grateful for the contribution of national delegations that provided information and commented on the draft. Other documents in the series cover scientific publishing, music, on-line computer games, mobile content, and public sector information and content (all available at **www.oecd.org/sti/digitalcontent**), and further reports are forthcoming on film and video and on-line advertising. Some of these reports are also summarised in the *OECD Information Technology Outlook 2006* (see **www.oecd.org/sti/ito**) or are forthcoming in the next edition, to be published in 2008.

Table of Contents

Foreword 3

Summary...**9**

Definition, measurement and drivers 9
Emerging value chains and business models 10
Economic impacts 11
Social impacts 12
Opportunities and challenges 13

Chapter 1. Introduction...**15**

**Chapter 2. Defining and Measuring the Participative Web and
User-Created Content**...**17**

Definition 17
Measurement 19

Chapter 3. Drivers of User-Created Content **27**

Technological drivers...27
Social drivers...29
Economic drivers 29
Institutional and legal drivers 30

Chapter 4. Types of User-Created Content and Distribution Platforms **31**

UCC types 34
Text..34
Photos and images 34
Music and audio 34
Video and film...35
User-created content posted on products and other interest areas 35

UCC platforms 36
  Blogs...36
  Wikis and other text-based collaboration formats 37
  Group-based aggregation and social bookmarking 37
  Podcasting 37
  Social networking sites...38
  Virtual world content..38

Chapter 5. Emerging Value Chains and Business Models **41**

The emerging value and publishing chain of user-created content 41
  Monetisation of user-created content and new business models 45

Chapter 6. Economic and Social Impacts **53**

Economic impacts 53
  Consumer electronics, ICT hardware, software, network service and platforms 55
  Users/creators 56
  Traditional media 57
  Professional content creators..61
  Search engines and advertising 61
  Services that capitalise on UCC 62
  Marketing and brands..62
  Use of UCC and participative web tools in business..63
Social impacts 63
  Increased user autonomy, participation and communication 64
  Cultural impacts 64
  Citizenship engagement and politics 65
  Educational and information impact..67
  Impact on ICT and other skills 68
  Social and legal challenges...68

Chapter 7. Opportunities and Challenges for Users, Business and Policy **71**

Enhancing R&D, innovation and technology...72
  R&D and innovation 72
  Ensuring technological and other spillovers...73
  Creative environments, skills, education and training..73
  Fostering local and diverse content 73
Developing competitive, non-discriminatory policy frameworks................................74
Enhancing the infrastructure 75
  Broadband access 75
  Convergence and regulation 76
Regulatory environment...77
  Intellectual property rights and user-created content..77
  Digital rights management 88

Freedom of expression 90
Information and content quality 90
Mature, inappropriate, and illegal content...92
Safety on the Internet and awareness raising...94
Privacy and identity theft...95
Impacts of intensive Internet use...96
Network security and spam 97
Virtual worlds, property rights and taxation..98
Governments as producers and users of content..99
Conceptualisation, classification and measurement 99

Annex..101

Bibliography 107

Notes 115

SUMMARY

The concept of the "participative web" is based on an Internet increasingly influenced by intelligent web services that empower users to contribute to developing, rating, collaborating and distributing Internet content and customising Internet applications. As the Internet is more embedded in people's lives users draw on new Internet applications to express themselves through "user-created content" (UCC).

This study describes the rapid growth of UCC, its increasing role in worldwide communication and draws out implications for policy. Questions addressed include: What is user-created content? What are its key drivers, its scope and different forms? What are new value chains and business models? What are the extent and form of social, cultural and economic opportunities and impacts? What are associated challenges? Is there a government role and what form could it take?

Definition, measurement and drivers

There is no widely accepted definition of user-created content and measuring its social, cultural and economic impacts are in the early stages. In this study UCC is defined as: *i)* content made publicly available over the Internet, *ii)* which reflects a certain amount of creative effort, and *iii)* which is created outside of professional routines and practices. Based on this definition a taxonomy of UCC types and hosting platforms is presented. While the measurement of UCC is in its infancy, available data show that broadband users produce and share content at a high rate, and this is particularly high for younger age groups (*e.g.* 50% of Korean Internet users report having a homepage and/or a blog). Given strong network effects a small number of platforms draw large amounts of traffic, and online video sites and social networking sites are becoming to be the most popular websites worldwide.

The study also identifies: technological drivers (*e.g.* more wide-spread broadband uptake, new web technologies), social drivers (*e.g.* demographic factors, attitudes towards privacy), economic drivers (*e.g.* increased commercial involvement of Internet and media firms in hosting UCC) and legal drivers (*e.g.* the rise of more flexible licensing schemes).

Emerging value chains and business models

Most user-created content activity is undertaken with no expectation of remuneration or profit. Motivating factors include connecting with peers, self-expression, and achieving a certain level of fame, notoriety or prestige. Defining an economic value chain for UCC as in the other OECD digital content studies is thus more difficult.

From a creator's point of view, the traditional media publishing value chain depends on various entities selecting, developing and distributing the creator's work often at great expense. Technical and content quality is guaranteed through the choice of the traditional media "gatekeepers". Compared to the potential supply, only a few works are distributed, for example, via television or other media.

In the UCC value chain, content is directly created and posted for or on UCC platforms using devices (*e.g.* digital cameras), software (video editing tools), UCC platforms and an Internet access provider. There are many active creators and a large supply of content that attract viewers, although of potentially lower or more diverse quality. Users are also inspired by, and build on, existing works as in the traditional media chain. Users select what does and does not work, for example, through recommending and rating, possibly leading to recognition of creators who would not be selected by traditional media publishers.

Most UCC sites have been start-ups or non-commercial ventures of enthusiasts, but commercial firms are now playing an increasing role in supporting, hosting, searching, aggregating, filtering and diffusing UCC. Most models are still in flux and revenue generation for content creators or commercial firms (*e.g.* media companies) is only now beginning. Different UCC types (*e.g.* blogs, video content) have different although similar approaches to monetising UCC. There are five basic models: *i)* voluntary contributions; *ii)* charging viewers for services, *e.g.* pay-per-item or subscription models, including bundling with existing subscriptions; *iii)* advertising-based models; *iv)* licensing of content and technology to third parties; and *v)* selling goods and services to the community ("monetising the audience via online sales"). These models can also remunerate creators, either by sharing revenues or by direct payments from other users.

Economic impacts

User-created content is already an important economic phenomenon despite it originally being largely non-commercial. The spread of UCC and the amount of attention devoted to it by users appears to be a significant disruptive force for how content is created and consumed and for traditional content suppliers. This disruption creates both opportunities and challenges for established market participants and their strategies.

The more immediate economic impacts in terms of growth, entry of new firms and employment are currently with ICT goods and services providers and newly forming UCC platforms. New digital content innovations seem to be more based on decentralised creativity, organisational innovation and new value-added models, which favour new entrants, and less on traditional scale advantages and large start-up investments. Search engines, portals and aggregators are also experimenting with business models that are often based on online advertisement and marketing. On social networking sites and in virtual worlds, for example, brands increasingly create special sub-sites, and new forms of advertising are emerging.

The shift to Internet-based media is only beginning to affect content publishers and broadcasters. At the outset, UCC may have been seen as competition as: *i)* users may create and watch UCC at the expense of traditional media, reducing advertising revenues; *ii)* users become more selective in their media consumption (especially younger age groups); *iii)* some UCC platforms host unauthorised content from media publishers. However, some traditional media organisations have shifted from creating on-line content to creating the facilities and frameworks for UCC creators to publish. They have also been making their websites and services more interactive through user comment and ratings and content diffusion. TV companies are also licensing content and extending on-air programmes and brands to UCC platforms.

There are also potentially growing impacts of UCC on independent or syndicated content producers. Professional photographers, graphic designers, free-lance journalists and similar professional categories providing pictures, news videos, articles or other content have started to face competition from freely provided amateur-created content.

Social impacts

The creation of content by users is often perceived as having major social implications. The Internet as a new creative outlet has altered the economics of information production, increased the democratisation of media production and led to changes in the nature of communication and social relationships (sometimes referred to as the "rise - or return - of the amateurs"). Changes in the way users produce, distribute, access and re-use information, knowledge and entertainment potentially give rise to increased user autonomy, increased participation and increased diversity. These changes may result in lower entry barriers, distribution and user costs and greater diversity of works, as digital shelf space is almost limitless.

UCC can provide citizens, consumers and students with information and knowledge. Educational UCC content tends to be collaborative and encourage sharing and joint production of information, ideas, opinions and knowledge, for example building on participative web technologies to improve the quality and extend the reach of education. Discussion fora and product reviews can lead to more informed user and consumer decisions (*e.g.* fora on health-related questions, book reviews).

The cultural impacts of this social phenomenon are also far-reaching. "Long tail" economics (the potential to distribute small quantities of products cheaply) allows a substantial increase in, and a more diverse array of, cultural content to find niche users. UCC can also be an open platform enriching political and societal debates and increasing diversity of opinion, the free flow of information and freedom of expression. Transparency and "watchdog" functions may be enhanced by decentralised approaches to content creation. Citizen journalism, for example, allows users to correct, influence or create news, potentially on similar terms to newspapers or other large entities. Furthermore, blogs, social networking sites and virtual worlds can be used for engaging electors, exchanging views, provoking debate and sharing information on societal and political questions.

Challenges related to exclusion, cultural fragmentation, content quality and security and privacy have been raised. A greater divide between digitally literate users and others may occur and cultural fragmentation may take place with greater individualisation of the cultural environment. Other challenges relate to information accuracy and quality (including inappropriate or illegal content) when everybody can contribute without detailed checks and balances. Other issues relate to privacy, safety on the Internet and possibly adverse impacts of intensive Internet use.

Opportunities and challenges

The rapid rise of UCC is raising new questions for users, business and policymakers. Digital content policy issues are grouped under six headings: *i)* enhancing R&D, innovation and technology; *ii)* developing a competitive, non-discriminatory policy framework; *iii)* enhancing the infrastructure; *iv)* shaping business and regulatory environments; *v)* governments as producers and users of content and *vi)* better measurement. Governments as producers and users are treated in more detail in separate work.

Apart from standard issues such as ensuring wide-spread broadband access and innovation, new questions emerge concerning whether and how governments should support UCC. The maintenance of pro-competitive markets is particularly important with increased commercial activity combined with strong network effects and potential for lock-in. UCC is also testing existing regulatory arrangements and the separation of broadcasting and telecommunications regulations. With the emergence of increasingly advertising-based business models and unsolicited e-mail and marketing messages, rules on advertising will play an important role in the UCC environment (*e.g.* product placements, advertising to children).

In the regulatory environment important questions relate to intellectual property rights and UCC: how to define "fair use" and other copyright exceptions, what are the effects of copyright on new sources of creativity, and how does IPR shape the coexistence of market and non-market creation and distribution of content. In addition, there are questions concerning the copyright liability of UCC platforms hosting potentially unauthorised content, and the impacts of digital rights management.

Other issues include: *i)* how to preserve the freedom of expression made possible by UCC; *ii)* information and content quality/accuracy and tools to improve these; *iii)* adult, inappropriate, and illegal content and self-regulatory (*e.g.* community standards) or technical solutions (*e.g.* filtering software); *iv)* safety on the "anonymous" Internet; *v)* dealing with new issues surrounding privacy and identity theft, *vi)* monitoring the impacts of intensive Internet use; *vii)* network security and spam, and *viii)* regulatory questions dealing with virtual worlds (taxation, competition etc.). Finally, new statistics and indicators are urgently needed to inform policy.

Chapter 1

INTRODUCTION

Wider participation in creating, distributing, accessing and using digital content is being driven by rapidly diffusing broadband access and easy-to-use software tools. Initial analysis of the participative web was published in the *Information Technology Outlook 2006*,[1] and heightened awareness of the growth and potential impacts of user-created content was a major outcome of the international conference on *The Future Digital Economy: Digital Content Creation, Distribution and Access* organised in Rome by the OECD and the Italian Minister for Innovation and Technologies in January 2006.[2]

As the Rome Digital Content conference progressed it became increasingly clear that the Internet is not only embedded in people's lives but that with the rise of a more "participative web" its impacts on all aspects of economic and social organisation are expanding (OECD, 2006a, 2006b). The conference increasingly focused on new "user" activities, where users draw on new Internet applications to create content and express themselves through "user-created content" and a more pro-active, collaborative role in content creation, distribution and use. More active users, consumers and user-centred innovation were seen to have increasing economic impacts and social importance. New forms of content creation and distribution are spurring new business models and are beginning to bypass, intersect with, and create new opportunities for, traditional media and content-related industries and access routes.

This study further expands published OECD work, exploring the development, rise and impacts of user-created content (UCC) in greater detail, and drawing out implications for policy. Questions addressed include: What is user-created content? What are its key drivers, its scope and what different forms does it take? What are new value chains and business models? What is the extent of its economic, social and cultural impacts? What are associated challenges? Is there a government role and, if there is, what form could it take?

The analysis in this report is divided into six main parts. The first part defines user-created content. The second and third parts identify the key drivers of UCC and provide a broad overview of various UCC types and related distribution platforms. The fourth part analyses associated "value" chains and new business models while the fifth part examines economic and social impacts. The final part analyses opportunities and challenges for users, business and government policy.

The development of user-created content is very recent and as general trends, impacts and related policies are still evolving, some questions raised in this report cannot be answered fully.

While open source software is often included as part of the participative web, it is not included in this analysis. However, in terms of economic and social impacts, the large-scale collaborative development and use of open-source software merits a great deal of further attention and analysis.

Chapter 2

DEFINING AND MEASURING THE PARTICIPATIVE WEB AND USER-CREATED CONTENT

Definition

The use of the Internet is characterised by increased participation and interaction of users to create, express themselves and communicate. The "participative web" is the most common term and underlying concept used to describe the more extensive use of the Internet's capabilities to expand creativity and communication. It is based on intelligent web services and new Internet-based software applications that enable users to collaborate and contribute to developing, extending, rating, commenting on and distributing digital content and developing and customising Internet applications (O'Reilly, 2002, 2005; MIC, 2006; OECD, 2006a, 2006b). New web software tools enable commercial and non-commercial service providers to draw on an ever-widening array of content sources and what is often called the "collective intelligence" of Internet users, to use information on the web in the form of data, metadata and user resources, and to create links between them. A further characteristic of the participative web is the communication between users and between separate software applications via open web standards and interfaces.

The rise of user-created content (UCC) (French: "contenu auto-créé") or the so-called "rise of the amateur creators" is one of the main features of the participative web but the participative web is a wider concept.[3] UCC comprises various forms of media and creative works (written, audio, visual, and combined) created by Internet and technology users. Despite frequent references to this subject no commonly agreed definition of user-created content exists.[4] Also referred to as "user-generated" content, sources such as Wikipedia refer to it as *"... various kinds of media content that are produced by end-users (as opposed to traditional media producers such as professional writers, publishers, journalists, licensed broadcasters and production companies)".*[5]

To have a more solid understanding of user-created content, three central characteristics are proposed below. These characteristics lay the ground for identifying a spectrum of UCC although they are likely to evolve over time.

- **Publication requirement**: A principle characteristic is that the work is *published* in some context, for example on a publicly accessible website or on a page on a social networking site only accessible to a select group of people (*e.g.* fellow university students), even though UCC could be made by a user and never published online or elsewhere. This characteristic excludes e-mail, two-way instant messages and the like.

- **Creative effort**: A *certain amount of creative effort has to be put into creating the work or adapting existing works to construct a new one*; *i.e.* users must add their own value to the work. UCC could include user uploads of original photographs, thoughts expressed in a blog or a new music video. The creative effort behind UCC may also be collaborative, for example on websites that users edit collaboratively. Merely copying a portion of a television show and posting it on an online video website (a frequent activity on UCC sites) would not be considered UCC. Nevertheless the minimum amount of creative effort is hard to define and depends on the context.

- **Creation outside of professional routines and practises**: User-created content is usually created *outside of professional routines and practices.* It often does not have an institutional or commercial market context and UCC may be produced by non-professionals without expectation of remuneration or profit. Motivating factors include: connecting with peers, achieving fame, notoriety or prestige, and expressing oneself.

Although conceptually useful it has become harder to maintain the last UCC characteristic of creators not expecting remuneration or profit and creation being outside of professional routines. UCC may have begun as a grassroots movement not focused on monetary rewards, but monetisation of UCC has been a growing trend (see section on economic impacts below). Established media and Internet businesses have increasingly acquired UCC platforms for commercial purposes. Some users are remunerated for their content and some become professionals after an initial phase of non-commercial activity. Some works are also created by professionals outside of their commercial activities (*e.g.* professional video editors creating a film at home). The term UCC may thus cover content creation by those who are much more than just "users". Still, the creation of content outside of a professional routine and organisation and potentially not for reward is a useful characteristic to separate it from content produced by commercial or quasi-commercial entities for commercial purposes.

Measurement

Measuring UCC is not straightforward. A number of factors complicate measurement: the decentralised nature of UCC production (sampling frames and defining the universe to be measured), the same UCC content may be accessible on different sites (issue of double-counting), not all registered users of UCC platforms may be active (inactive accounts), users may set up multiple accounts at the same site (identification of unique users), and distinguishing between user-created and other content (such as the uploading of clips from copyrighted television shows). The third and fourth factors may lead UCC platforms to overestimate the number of currently active unique users.

Little official data is available from National Statistical Offices (NSOs) on the numbers of users creating content, the amount that exists, the numbers of users accessing such content and economic and social patterns emerging from such creation. NSOs have only recently started to include such questions in surveys (*e.g.* Canada, the European Union, Japan, Korea). It will take some time before official national data is available for all OECD countries in an internationally comparable forum.

Existing data however show that broadband Internet users produce and share content at a high rate and do not merely consume it, and all data sources point to large intergenerational differences in web media usage and to considerable gender differences in usage.

Data available from national statistical surveys and the OECD show that the typical online behaviour of Internet users mainly consists of: searching, consulting general interest sites and portals, using Internet tools and web services such as e-mail, e-commerce, using sites from software manufacturers, consulting classifieds and participating in auctions, using broadcast media, and financial services (OECD, 2004a; OECD, 2005a).

Available data shows that content creation is a very popular activity among young age groups. As shown in Figure 2.1 for the European Union, statistical proxies that measure some aspects of UCC –- posting messages to chat rooms, newsgroups or forums, using peer-to-peer file sharing sites and creating a webpage –- are already very popular among Internet users.[6] In Finland, Norway, Iceland, Portugal, Luxembourg, Hungary and Poland (in increasing order), around one third of all Internet users aged 16-74 were engaged in one of these activities in 2005, most commonly posting messages in all countries. One-fifth of all Internet users in a few OECD countries report having created a webpage. Younger age groups are more active Internet content creators. In Hungary, Denmark, Iceland, Finland, Norway, Germany, Poland and Luxembourg (in increasing order), in 2005 between

Figure 2.1a. User-created content creators in the EU as a % of Internet users, 2005

Age group: 16-24 years

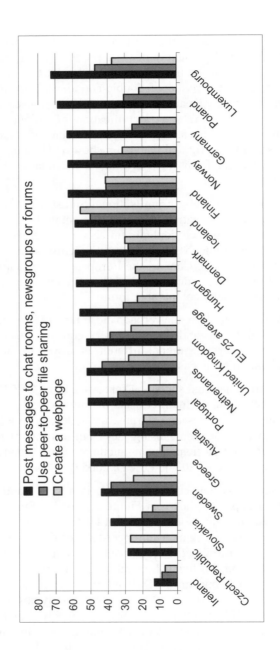

Figure 2.1b. User-created content creators in the EU as a % of Internet users, 2005

Age group: 16-74 years

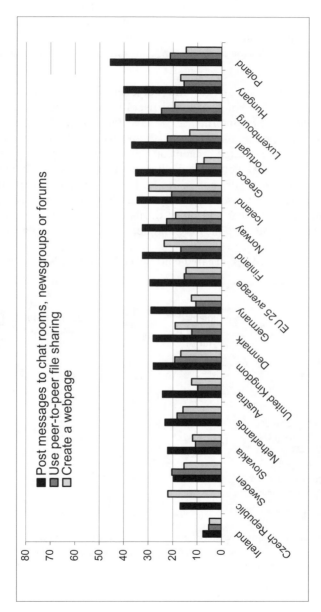

Source: OECD based on Eurostat.

from 55% to over 70% of Internet users aged 16-24 posted messages to chat rooms, newsgroups or forums. One quarter but sometimes half of all Internet users in some OECD countries in that age group have created a web page. In France, about 37% of teenagers have created a blog.[7] In 2005, 13% of Europeans were "regularly contributing to blogs" and another 12% were "downloading podcasts at least once a month" (European Commission, 2006).

Over one-third of all US Internet users have posted content to the Internet (Table 2.1). For broadband users under the age of 30, 51% have placed content on the Internet, 25% have their own blogs, and 41% have posted content online they created themselves. 57% of teenagers in the United States have created content on the Internet as of late 2004 (Lenhart, 2005). Online social networking sites were used by more than one-half (55%) of all online Americans in the 12-17 age group (Lenhart and Madden, 2007). In general, girls seem to use social networking sites relatively more for communication, chat and other forms of socialising and exchange, but less for just viewing content, for example, on online video platforms.[8]

Table 2.1. User-created content in the United States, 2006

	All Internet users (in %)	Broadband at home (in %)	Americans with home Internet access who do the activity in question (in millions)	Americans with Internet access only at places other than home or work (in millions)
Create or work on your own online journal or blog	8	11	9	2
Create or work on your own web page	14	17	18	2
Create or work on web pages or blogs for others including friends, groups belonged to, or workmates	13	16	16	2
Share something online that you created yourself, such as your own artwork, photos, stories or videos	26	32	32	4
Percentage having performed at least one of the above "content" activities	35	42	43	5

Source: OECD based on PEW Internet & American Life Project, December 2005 survey.

Note: Margin of error for Internet users is +/- 2%. See also the presentation of John B. Horrigan (Pew Internet & American Life Project) to the OECD.[9]

Official data from Asian countries show similar user behaviour. Data for Japan show that blogs and social networking sites (SNS) have substantially increased the amount of information available on the Internet (MIC, 2006). At the end of March 2006, 8.7 million Japanese were registered as bloggers, and 7.2 million as SNS members (see Figure 2.2). About one quarter of Japanese Internet users over 12 years of age have experience in finding friends/acquaintances through the Internet and close to one fifth have had exchanges with people they did not previously know. The 20-30 age group has the most experience in finding friends and acquaintances online, with women more active than men. Of those with experience in such online exchanges, almost 50% subsequently met their online acquaintances offline.[10]

A recent Korean Internet use survey shows that about 50% of Korean Internet users use the Internet for managing homepages and/or blogs. In China, around 43% of all Internet users use electronic bulletin boards, online communities and fora and instant messengers and 24% use a blog (CNNIC, 2006).

Figure 2.2. Number of registered blog and social networking site users in Japan, in millions

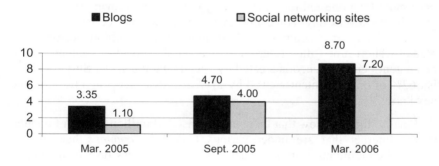

Source: U-Japan presentation at www.apectel34.org.nz/uploads/061016%20Kondo.pdf.

New measurement tools which track Internet audience and traffic provide insights into the development of UCC. Private data providers are tracking millions of users and their Internet activity anonymously (large sample sizes with real-time flow of information and no need to gather data via surveys) and knowledge about Internet usage is growing rapidly. Firms which measure Internet traffic such as Hitwise, comScore or Nielsen NetRating show the increasing attraction of sites hosting UCC.[11] For instance, during 2006 the five UCC sites that ranked in the top 50 sites in the

United Kingdom (measured by total visitors that month) — Wikipedia, MySpace, Piczo, YouTube and Bebo — generated an average of 4.2 usage days and 79.9 minutes per visitor, according to comScore. By comparison, sites in the top 50 that were not based on UCC saw far less frequent usage days and times.

These UCC sites have also experienced significant traffic growth. According to Internet traffic data of Alexa, YouTube continues to grow and the online video site is now ranked the number four web site in the world - behind only Yahoo, MSN and Google itself.[12] A study by Nielsen NetRating shows that in the United Kingdom UCC platforms for photo sharing, video sharing and blogging are among the fastest growing Web sites. In the US, UCC sites comprised five out of the top 10 fastest growing web sites in July 2006. Among the top 10 web sites overall, MySpace was the No. 1 fastest growing, increasing to 46 million unique visitors in July 2006 (see Table 2.2). In countries such as Korea, UCC has contributed to reinvigorating the growth of some web sites. Given the strong network effects which are a characteristic of UCC sites, a small number of platforms draw large amounts of traffic (*i.e.* a high concentration of users on only a few platforms).

Internet measurement services created to measure the size of particular content types (such as Technorati for blogs) and data from the sites hosting UCC, also point to this fast growth. Popular video sharing sites, for instance, serve more than one million videos every day and more than 50 000 videos are uploaded every day. Anecdotal evidence shows that certain videos are shared widely, some being viewed by more than 1 million persons in a relatively short time (often referred to as a "viral" spread of content). Often the content being watched is created by friends, families, etc. Surveys claim that 62% of online content viewed by 21-year-olds is generated by someone they know.[13]

Finally, figures show that there are at least 200 million pieces of content on the Internet that are under various *Creative Commons* licences (as counted by the number of "link-backs" to these licences on the Internet as tracked by Google).

All of these official and unofficial data show the rapid rise and increasing use of user-created content.

Table 2.2. Fastest growing web sites among US at-home and at-work Internet users, July 2006

Firm	Type of Internet service	Number of unique visitors (in millions)	% growth of unique users from July 2005 to July 2006
HSBC	Bank	6.4	394
Sonic Solutions	Provider of digital media software	3.7	241
Associated Press	Press agency	9.7	234
ImageShack	Image hosting site	7.7	233
Heavy.com	Video sharing site	3	213
Flickr	Photo sharing site	6.3	201
ARTIST Direct	Online music platform including for emerging artists	3.2	185
Partypoker.com	Online gambling site	6	184
MySpace	Social network site	46	183
Wikipedia	Online community project	29.2	181

Source: Nielsen//NetRatings, July 2006, www.nielsen-netratings.com/pr/PR_060810.PDF. Shading indicates sites relevant to UCC. The term unique visitor refers to a person who visits a website more than once within a specified period of time.

Chapter 3

DRIVERS OF USER-CREATED CONTENT

There a range of technological, social, economic and institutional drivers of user-created content accounting for its rapid growth and pervasiveness. These are summarised below and in Box 3.1.

Technological drivers

First, the rapid uptake of broadband by households from the late 1990s has increasingly enabled users to create, post and download content. The limitations of dialup connections meant that user content creation was largely restricted to text and simple, low quality graphics. However, with high speed connections, users could quickly create and upload ever-larger media files. The ease of creation, uploading and downloading will amplify as fibre to the home/premises becomes widespread,[14] as high-speed wireless broadband becomes available and as newer-generation ubiquitous networks spread.

Second, there have been large increases in processing speeds, hard drive and flash memory capacities and consumer electronics capabilities (high quality digital cameras, digital video recorders and mobile phones) to create content, while their price/performance ratios have decreased sharply. New mobile phone platforms with High-Speed Uplink Packet Access (HSUPA) and higher uplink data transmission speeds will further enable users to send and receive mobile phone clips and pictures at higher speeds.

Third, more accessible software tools, such as html-generating software, and software which enables users to find, edit and create audio and video without professional knowledge are another driving force. Because UCC is posted widely on the Internet, the challenge of locating, distributing, and assessing the quality of the content has spurred other new technologies which facilitate tagging (*i.e.* the association of particular keywords with related content), podcasting, group rating and aggregation, recommendations, content distribution (*e.g.* Really Simple Syndication, RSS, feeds which ensure that users automatically receive new posts and updates and file-sharing software), and technologies for interactive web applications, filtering and feeding such as Ajax, RSS, Atom and other content management systems necessary for blogs, wikis and other content (see Annex Box 1 and OECD, 2006a).

Box 3.1. Drivers of user-created content

Technological drivers

Increased broadband availability

Increased hard drive capacity and processing speeds coupled with lower costs

Decrease in cost and increase in quality of consumer technology devices for audio, photo, and video

Availability of technologies to create, distribute, and share content

Development of simpler software tools for creating, editing, and remixing

Rise of non-professional and professional UCC sites as outlets

Social drivers

Shift to younger age groups ("digital natives") with substantial ICT skills, willingness to engage online (*i.e.* sharing content, recommending and rating content, etc.) and less hesitant to reveal personal information online

Desire to create and express oneself and search for more interactivity than on traditional media platforms such as TV

Development of communities and collaborative projects

Spread of social drivers to older age groups and for societal functions (social engagement, politics and education)

Economic drivers

Lower costs and increased availability of tools for the creation of UCC (*e.g.* for creating, editing, hosting content) and lower entry barriers

Lower cost of broadband Internet connections for providers and users

Increased commercial interest in user-created content and "long tail" economics (including mobile operators, telecommunication service providers, traditional media publishers and search engines)

Increased possibilities to finance UCC-related ventures and sites through venture capital and other investment vehicles

Greater availability of advertising and new business models to monetise content

Institutional and legal drivers

Rise of schemes which provide more flexible access to creative works and the right to create derivative works — (*e.g.* flexible licensing and copyright schemes such as the Creative Commons licence)[15]

Rise of end-user licensing agreements which grant copyright to users for their content.

Finally, the rise of sites and services hosting UCC was a necessary driver as not every user has available server space or the technical skills to post and distribute their work. As the quality of cameras and video capabilities on phones grows and as phone networks are increasingly integrated with the Internet, mobile content (*e.g.* mobile blogging) is spreading more widely, and depends on new mobile services. New video platforms that feature UCC such as IPTV services (*e.g.* transmission of TV programming over broadband using peer-to-peer technology and technologies allowing high-resolution broadband video transmission), and video game consoles geared to UCC will provide additional impetus.

Social drivers

Increased use of broadband, greater on-line interactivity and the willingness to share, contribute and create online communities are changing media consumption habits of Internet users, in particular younger age groups. Social factors are likely to be one of the most important drivers of change. UCC is only starting to move mainstream, with initially a limited number of young, male early adopters and highly ICT-skilled persons using the Internet in this way. According to surveys, almost three-quarters of people who publish amateur video content online are under 25, and of those, 86% are male.[16] Overall, user-created video is viewed by a large number of people but created by only a few.

Economic drivers

There is also increased interest in monetising UCC. Media companies, the communications industry (in particular mobile operators), and other commercial operators have identified the potential of UCC and are investing substantially in new and established UCC ventures. Slow-downs in revenues due to decreased interest in traditional media and the desire to cater to the so-called "long tail" (the potential to distribute small quantities of products cheaply) have been important driving factors (Anderson, 2004). This interest is also reflected in the growing amount of private and corporate financing and venture capital available for investment in UCC related sites and services. In the United States, for example, venture capital funding related to the participative web Internet technologies were estimated to have increased by more than 40% from the third quarter of 2005 to the third quarter of 2006, and venture capital investments in information services companies (covering Web 2.0 internet companies that run social networks, blogs and wikis as well as IT-based services such as database design) were USD 979 million in the second quarter of 2007, 52% higher than in the second quarter of 2006.[17] While significant, however, total venture capital invested in ICT and media are still only about a quarter of the investments at the height of the boom in 2000.

Institutional and legal drivers

The rise of new legal means to create and distribute content has also contributed to the greater availability and diffusion of UCC. Flexible licensing and copyright schemes such as the Creative Commons licences allow easier distribution, copying and – depending on the choice of the author – the creation of derivative works of UCC.[18] Increasingly search engines and UCC platforms allow for searches within Creative Commons-licensed photos, videos or other content allowing others to use and build on them. The rise of end-user licensing agreements (*e.g.* Second Life) which grant copyright to users for the content that they create may also be a significant driver.

Chapter 4

TYPES OF USER-CREATED CONTENT AND DISTRIBUTION PLATFORMS

A range of different types and distribution platforms for user-created content have developed, with a significant amount relying on hosting services providing online space where the content can be accessed. This section gives an overview of common types of UCC and UCC distribution platforms (see Kolbitsch and Maurer, 2006 and Tables 4.1 and 4.2).

Different types of UCC types are often linked to specific UCC distribution platforms, *e.g.* written comments being diffused on blogs, videos being diffused on online sharing platforms, and UCC types and their distribution platforms are often closely associated (Tables 4.1 and 4.2). However, some UCC distribution platforms such as podcasting are used for music and video with various purposes (entertainment, educational, etc.), and social networking sites can be used to post music, videos, to blog, etc.

Moreover, participative web technologies often originally used for UCC can also be used for traditional media, other commercial or educational content (*e.g.* podcasts of well-known news magazines, games or social networking site used for commercial or educational content). Businesses may also make use of weblogs to keep employees informed of new products and strategies or on the progress of projects and for other internal communications, but such activities are not discussed in detail in this study.

The following sections describe selected UCC types and distribution platforms. Some UCC types such as video are described only once even though they may appear on a range of UCC platforms.

Table 4.1. Types of user-created content

Type of content	Description	Examples
Text, fiction and poetry	Original writings or expanding other texts, novels, poems	Fanfiction.net, Quizilla.com, Writely
Photos and images	Digital photographs taken by users and posted online; photos or images created or modified by users	Photos posted on sites such as Ofoto and Flickr; photo blogging; remixed images
Music and audio	Recording and/or editing personal audio content and publishing, syndicating, and/or distributing in digital format	Audio mash-ups, remixes, home-recorded music on band websites or MySpace pages, podcasting.
Video and film	Recording and/or editing video content and posting it. Includes remixes of existing content, homemade content, and combinations of the two.	Movie trailer remixes; lip-synching videos; video blogs and videocasting; posting home videos. Sites include YouTube and Google Video; Current TV
Citizen journalism	Journalistic reporting on current events by ordinary citizens who write news stories, blog posts, and take photos or videos of current events and post them online.	Sites such as OhmyNews, GlobalVoices and NowPublic; photos and videos of newsworthy events; blog posts reporting an event; co-operative efforts such as CNN Exchange
Educational content	Content created in schools, universities, or for educational use	Syllabus-sharing sites such as H20; Wikibooks, MIT's OpenCourseWare
Mobile content	Content created on mobile phones or other wireless devices such as text messaging, photos and videos. Generally sent to other users via MMS (Media Messaging Service), e-mailed, or uploaded to the Internet.	Videos and photos of public events or natural catastrophes that traditional media may not be able to cover; text messages for political rallying.
Virtual content	Content created within the context of an online virtual environment or integrated into it. Some virtual worlds allow content to be sold. User-created games.	Virtual goods that can be developed and sold on Second Life including clothes, houses, artwork

Table 4.2. Distribution platforms for user-created content

Type of platform	Description	Examples
Blogs	Web pages containing user-created entries updated at regular intervals and/or user-submitted content investigated outside of traditional media	Popular blogs such as BoingBoing and Engadget; blogs on sites such as LiveJournal; MSN Spaces; CyWorld; Skyblog
Wikis and other text-based collaboration formats	A wiki is a website that allows users to add, remove, or otherwise edit and change content collectively. Other sites allow users to log in and co-operate on the editing of particular documents.	Wikipedia; sites providing wikis such as PBWiki, JotSpot, SocialText; writing collaboration sites such as Writely
Sites allowing feedback on written works	Sites which provide writers and readers with a place to post and read stories, review stories and to communicate with other authors and readers through forums and chat rooms	FanFiction.Net
Group-based aggregation	Collecting links of online content and rating, tagging, and otherwise aggregating them collaboratively	Sites where users contribute links and rate them such as Digg; sites where users post tagged bookmarks such as del.icio.us
Podcasting	A podcast is a multimedia file distributed over the Internet using syndication feeds, for playback on mobile devices and personal computers	iTunes, FeedBruner, iPodderX, WinAmp, @Podder
Social network sites	Sites allowing the creation of personal profiles	MySpace, Facebook, Friendster, Bebo, Orkut, Cyworld
Virtual worlds	Online virtual environment	Second Life, Active Worlds, Entropia Universe, Dotsoul Cyberpark
Content or filesharing sites	Legitimate sites that help share content between users and artists	Digital Media Project

Note: Podcasting, blogs and related technologies are also increasingly used professionally (see OECD, 2006a for more discussion).

UCC types

Text

Users create texts, poems, novels, quizzes and jokes and share them with their communities. This allows the spread of works of amateur authors and community feedback. Fan fiction is often used to describe creative writing (often short stories) using pre-existing characters from television, movies or other fiction. Fanfiction.net is a fan site with thousands of stories for example expanding on J. K. Rowling's characters in Harry Potter books. Quizilla.com is an online, user-creative community of original teen authors who create and share quizzes, fiction, non-fiction, poetry, etc. Writing collaboration sites such as Writely support collaborative work on texts.

Photos and images

User-created photos are generally taken with digital cameras. Photos may or may not be manipulated with photo editing software. Advances in the aggregation and search functionalities via tagging, user-implemented indicators, and recognition software have changed the landscape of digital photos. Content on some sites is largely published under a Creative Commons licence, building an attractive resource for web designers, publishers and journalists. There are numerous services that have evolved around the hosting of photos, including Flickr, Ofoto,[19] and Snapfish. In 2006 the popular photo sharing service Flickr hosted 200 million photos taken by 4 million users, 80% of which were available to the general public.

Music and audio

User-created audio content on the Internet varies widely, ranging from the combination of two or more songs into a single track to the posting of self-created music by amateur musicians to creating a radio-like broadcast show that users can subscribe to (*i.e.* podcasts). Audio content may be hosted on sites dedicated to remixing, on sites that provide podcasting services, traded on peer-to-peer networks, posted on social networking sites, and on personal homepages and websites. So far, user-created music has rarely been listed by digital music stores. While there is a significant amount of user produced or recorded music posted on the Internet, remixes have gained notoriety. Remixing is however common in various genres of music, including hip hop and electronic, and in professional contexts.[20] Artists such as David Bowie have encouraged users to mash-up their music (OECD, 2005b).[21]

Video and film

User-produced or edited video content has taken three primary forms: homemade content, such as home videos or short documentaries; remixes of pre-existing works such as film trailer remixes; and hybrid forms that combine some form of self-produced video with pre-existing content. Examples include Chinese teenagers lip-synching (see Figure 4.1). Another type of user-created video consists of splicing up portions of videos or movies and creating new versions, often perceived as mock "trailers" for one or more of the films involved. Examples includes the various mash-up "spoofs" (*e.g.* parody by imitation) surrounding the film Brokeback Mountain.[22] Popular videos may also spur waves of remixes. Creators may use this form of remixing as social, political and cultural parody.

Video content may be hosted on a user's website, traded on peer-to-peer networks, private web pages or hosted by video sharing platforms such as YouTube, Google Video, AOL Uncut, Guba, Grouper and vPod in Europe, Dailymotion in France, MyVideo and Sevenload in Germany and, in Italy, Libero Video (see Annex Box 2 on China). Increasingly these sites are also enabled for access (upload and download) from mobile phones and devices. Stickam.com (live broadcasts from web cameras) and LiveLeak (reality-based footage) are among increasingly unfiltered video services.

Figure 4.1. Example of a lip-synching video

Source: YouTube.

User-created content posted on products and other interest areas

A large and heterogeneous category comprises users and consumers posting opinions and advice, also referred to as information and knowledge commons (called "word-of-mouth" sites in MIC, 2006). These take the form of Internet-based bulletin boards where contributors can submit opinions and critiques, *e.g.* product reviews. Other users, in turn, can use this information to make informed purchase decisions,[23] and businesses can more easily find out what the consumers feel about their products.[24] Ideas for new products or modifications can be gathered (see also the concept of user-led innovation in van Hippel, 2005).

Topics discussed are not limited to product reviews. Internet platforms (*e.g.* blogs) are used to exchange or present a wide range of knowledge or information, ranging from housing, health, computer problems, financial investments,[25] and travel advice to hobbies.[26] Some sites allow questions to be potentially answered by other users (*e.g.* the Yahoo "Ask a question" service). Many users find the Internet and community sites very useful, with targeted information/knowledge with a significant personal touch.

UCC platforms

Blogs

A blog is defined as a type of webpage usually displaying date-stamped entries in reverse chronological order (Gill, 2004; OECD, 2006a). It is updated at regular intervals and may consist of text, images, audio, video, or a combination of them. Blogs serve several purposes including delivering and/or sharing information. Installing blogging software – *e.g.* Movable Type, WordPress and Nucleus CMS – on a server is necessary to blog. However, blog hosting services (*e.g.* Blogger) make it easier by removing the technical burden of maintaining a hosting account and a software application. Often blogs are a launch pad for sharing other kinds of UCC, *i.e.* blogs typically refer to other blogs, music or discuss user-created videos. In 2007, video blogging is expected to grow very significantly.

Some sources estimate that there were up to 200 million blogs in 2006 (Blog Herald); the blog tracking site Technorati tracked 55 million blogs in December 2006 and estimated that they had doubled approximately every 6 months over the previous two years.[27] An approximation of the language distribution shows that nearly 75% of all blogs are written in English, Japanese or Korean.[28] Blogging is also very popular in China, India, and Iran. Their popularity in Asia is also shown in a recent Microsoft survey which suggest that nearly half of all Asian Internet users have a blog, that young users are most prevalent (56% of all bloggers are under 25, while 35% are 25 to 34 years old, and 9% are 35 years old and over) and that women are very active (55% of bloggers). Blogging is considered a form of expression and a means to maintain and build social connections (74% find blogs by friends and family most interesting).[29]

Wikis and other text-based collaboration formats

A wiki is a website that allows users to add, remove and otherwise edit and change content (usually text). Users can change the content of pages and format them with a very simple tagging language. Initial authors of articles allow other users to edit "their" content. The fundamental idea behind wikis is that a large number of users read and edit the content, potentially enriching it and correcting mistakes.

Various sites provide wiki hosting. Sometimes termed "wiki farms", these sites enable users and communities to create their own wiki for various purposes. In addition, forms of collaborative writing have developed with wiki technology (*e.g.* Writely, owned by Google, and Writeboard).[30] One frequently cited example is the online encyclopaedia Wikipedia. It comprised 4.6 million articles in over 200 languages in 2006 (Wikipedia, 2006). Fifteen of these languages had over 50 000 articles, with 1.3 million articles in English. The vast majority of edits come from a small percentage of users (Annex Table 1).

Group-based aggregation and social bookmarking

This is relatively new and consists primarily of group-based collection of links to articles and media and/or group based rating of such links, also referred to as new social content aggregators, which build on opinions and knowledge of all web users. Users generally collect these links, tag them, rate them, and often comment on the associated article or media. Sites such as Digg specialise in the use of this model, whereby users post news links to the site, and other users rate them by adding their vote to it.[31] Del.icio.us, a social bookmarking website, allows users to post links to their favourite articles, blogs, music, recipes, and more.

Podcasting

Podcasting has emerged out of the combination of the ease of audio production with technologies that allow for subscription and syndication. The publish/subscribe model of podcasting is a version of push technology, in that the information provider chooses which files to offer in a feed and the subscriber chooses among available feed channels. A consumer uses an aggregator software, sometimes called a podcatcher or podcast receiver, to subscribe to and manage feeds. Well-known podcast software includes FeedBurner, iPodderX, WinAmp and @Podder. Mobile-casting, *i.e.* receiving video and audio podcasts on mobile phones is expected to develop rapidly.

Podcasting technology is also used for content which does not come directly from users. Surveys estimated that 6 million Americans had listened to podcasts by 2005 (Raine and Madden, 2005). Popular download sites such as Apple iTunes hosted almost 83 000 podcasts in March 2006 (up from 8 000 one year earlier – see Annex Table 2 for the top categories).

Social networking sites

Social networking sites (SNS) enable users to connect to friends and colleagues, to send mails and instant messages, to blog, to meet new people and to post personal information profiles. Profiles include photos, video, images, audio, and blogs. In 2006, MySpace had over 100 million users (although not all are active) and is the most popular website in the United Sates according to Hitwise. Other popular SNS include Friendster, Orkut and Bebo. Facebook is popular on US college campuses with over 9 million users. The Korean Cyworld is reported to have 18 million users, or 40% of the population and 90% of Internet users in their 20s (Jung-a, 2006). Mixi, a Japanese SNS, has more than 4 million users.[32] Some video sharing sites such as Grouper allow private video sharing, furthering the social network dimension.

Some SNS sites are dedicated to particular topics, sharing knowledge, or purchases of products and services, transforming, for example, how users research, search and decide on travel plans. Yahoo's Trip Planner, Google's Co-Op, TripAdvisor's Inside, VirtualTourist's Trip Planner and others share travel journals, itineraries and photos. Similar social networking tools are used for real estate purchases.

Virtual world content

Virtual world content is created in an online game-like 3D digital environment to which users subscribe, although not all online multiplayer games allow users to create their own content. Virtual environments such as those in Second Life, Active Worlds, Entropia Universe, and Dotsoul Cyberpark provide users with a scripting language and integrated development environment which enables them to build new objects (Mayer-Schoenberger and Crowley, 2005), often permitting them to keep the associated intellectual property rights (see Figure 4.2. for an exhibition in Second Life).[33]

Figure 4.2. Library of Congress exhibit in Second Life

Source: Flickr.com.

In January 2007, Second Life claimed over 880 000 users in more than 90 countries who had logged on in the previous 60 days (and 2.5 million total residents).[34] Owning land in Second Life allows users to build, display, and store virtual creations, as well as host events and businesses or real university courses. It has an economy based on so-called Linden Dollars with more than USD 130 million per year contracted between players. Users can make money by selling created items (*e.g.* clothes for avatars) and land purchased earlier.

Chapter 5

EMERGING VALUE CHAINS AND BUSINESS MODELS

New value chains and business models are developing around user-created content and an increasing range of commercial and non-commercial participants are involved.

The emerging value and publishing chain of user-created content

UCC by definition is for the most part a non-commercial phenomenon and the vast majority of user-created content is created without the expectation of profit or remuneration. Motivating factors include connecting with peers, achieving fame, notoriety or prestige, and expressing oneself. Defining a value-chain for UCC in the traditional, commercial sense as for other OECD studies of digital content sectors is thus less straightforward. This analysis of business models and value chains is therefore a snapshot of emerging approaches which may have to be revisited with the development of UCC.

The value chain and distribution model for UCC is compared with a simplified established offline media publishing value chain (see Figure 5.1). This comparison mainly applies for content such as text, music, movies and similar media but is less applicable for content created in virtual worlds.

From a creator's point of view, the traditional media publishing value chain is characterised by a number of stages, *i.e.* publication and distribution of content depend on various entities selecting and consenting to a creator's work. To produce and publish their work, an individual writing editorials has to be recruited by a newspaper, a musician has to sign a record deal, a poet has to find a book publisher and a film writer has to successfully submit his script to film studios.

Figure 5.1. Traditional offline media publishing value chain

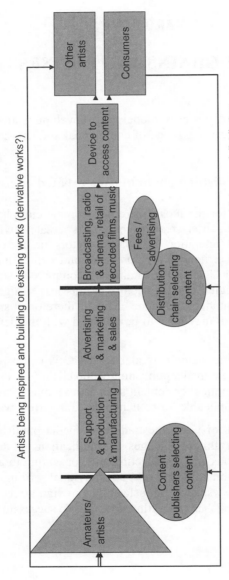

Figure 5.2. Original Internet value chain for user-created content

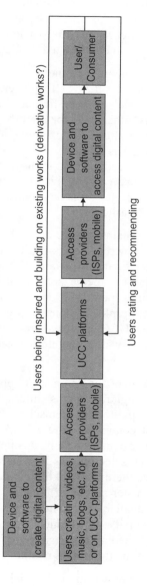

Users being inspired and building on existing works (derivative works?)

Users rating and recommending

Table 5.1. Selected recent acquisitions of UCC platforms

Date	Acquirer	Acquired	Type	Price (USD millions)
Sept. 2005	News Corporation	MySpace	Social networking site	580
Oct. 2005	Viacom/MTV	iFilm	Video	49
Aug. 2006	Sony	Grouper	Video	65
Aug. 2006	Viacom/MTV	Atom Films	Games, films, animation	200
Sept. 2006	Yahoo	Jumpcut	Video editing	Undisclosed
Oct. 2006	Viacom/MTV	Quizilla.com	Text, quizzes, images	Undisclosed
Oct. 2006	Google	YouTube	Video	1580
Nov. 2006	Google	Jotspot	Wiki	Undisclosed

Source: Company information and press reports.

The creation of the work to be published and its physical distribution can be, and usually are expensive. The content publishers select content, support the creator in production, manufacturing and advertising and in selecting the best distribution chain. Technical quality and content quality (although content quality is hard to define) is guaranteed by traditional media "gatekeepers". Compared to the potential supply, only a few works make it through to airing on television, radio, being distributed on CD, etc.

Consumers then watch the content on television, read the book, or listen to the CD using the appropriate media devices (*e.g.* CD player, radio receiver), and the number of distribution channels may be limited for some platforms, *e.g.* for television. The content is either paid for directly by the consumer (*i.e.* purchase of the media or subscriptions, for example, to cable TV) or is brought to the consumer on the basis of advertising-supported distribution channels. Customer preferences feed back into how content publishers, distributors and artists select future content. Finally, available works influence creation of new works, *i.e.* jazz players, playwrights, scriptwriters, singers are inspired by works of earlier artists.

In contrast, Figure 5.2 depicts the value chain for user-created content. The content is initially provided on non-commercial terms by its creators, potentially seeking recognition, fame or later financial reward. While UCC may rarely be a perfect substitute for traditional media content, it creates value for its viewers; as evidenced by the time spent by users downloading and watching (i.e. potentially high consumer surplus as content is generally free). Although the content itself may be free, UCC creates a strong demand for commercial products such as devices, software and Internet access to create and consume the content.

In the UCC value chain users create content for or on UCC platforms while using content creation devices (*e.g.* digital cameras, microphones), software (video editing tools), the UCC platforms themselves and an Internet access provider to create and post content (Figure 5.2). All users with access are able to create and publish their content widely, as opposed to the traditional media publishing chain and its selectivity as to what content shall be published. In some cases, users have their personal blog which does not rely on an external UCC platform. A greater supply of creative content from a larger number of active creators is available and engaging viewers, even if potentially of lower or greater diversity of technical and content quality. Similarly to artists in the traditional media chain, users are inspired and build on existing works – including from the traditional media.

The users themselves select which content works and which content does not, through recommending and rating (*i.e.* another form of advertising), possibly leading some creators to recognition and fame which would not have been given by traditional media publishers. The time it takes for content to be created and distributed is greatly reduced compared with the traditional value chain, which may impact the type and quality of content in multiple ways.

Users access the content by downloading or streaming it from the UCC platform through their access provider, using devices (*e.g.* notebook computer) and software (*e.g.* video streaming software).

Monetisation of user-created content and new business models

Commercial entities, including media companies, play an increasing role in supporting, searching, aggregating, filtering, hosting, and diffusing UCC. Direct revenue generation for the content creators or for established commercial entities (*e.g.* media companies, platforms hosting UCC) has only recently begun.

Until recently, sites hosting UCC were essentially non-commercial ventures of enthusiasts, or start-ups with little or no revenue but with increasing finance from venture capitalists. These sites often did not have business plans showing how revenues would be produced, rather their objective was to expand their user/creator and audience base, with an eye to either sell their business or implement paying business models at a later stage. However, despite low or non-existent revenues, considerable financial resources are necessary for the technology, bandwidth and organisation to keep operating, and given that many UCC platforms host unauthorised third party content they face demands for remuneration of content originators.

UCC sites are of increasing investor and business interest. For example, Mixi, the Japanese SNS site, and Open BC/Xing, a German business SNS site, have been listed on stock exchanges. Moreover, established media conglomerates and Internet companies are increasingly interested, with firms such as News Corporation, Google, Sony and Yahoo spending significant amounts to buy UCC sites (see Table 5.1).[35]

The increasing amounts being paid for UCC sites and the increased venture capital flowing into these areas have triggered renewed concerns about the build-up of a new Internet bubble. As in the late 1990s, the size of the website audience / "user engagement" ("eyeballs", traffic and page views and click-throughs) are drawing investors' attention. Earnings and revenues do not seem to be the prime concern. The large sums invested in buying up UCC start-ups have raised concerns of a second Internet bubble. While this cannot be excluded, in some respects the environment for these investments has changed with new possibilities associated with online advertising, new possibilities to deliver high-quality content through broadband, changed usage habits, increased ICT skills, etc. Furthermore, the overall sum of venture capital flowing into ICT-related areas in 2006 was still relatively small, only about 40% of average investments between 1999 and 2001 in the United States.

New models are developing on both the host- and creator-side of UCC spurred by an increased interest in monetising UCC. While the UCC value chain (*i.e.* the entities and activities that add value in producing and distributing the content) remains largely unchanged (see Figure 5.2), new models aim at the monetisation of this content. At the point when consumers access the UCC platform or a particular video, they donate, pay fees or subscribe to access the content or they see online advertising. New interactions with the established media value chain are emerging as UCC platforms are screened for promising talent and content which are later aired or integrated in the traditional media publishing value chain (*e.g.* in existing cable or TV subscriptions that are already subscribed to).

The advertising industry, search engines, and media firms who own UCC platforms or who select content from them are increasingly involved in the provision and distribution of content. When payments are involved, financial service providers and the associated technologies enter the value chain. As increasingly there is a need for tools to find content (*e.g.* search engines adapted to music, video and other multimedia content and user ratings and recommendations), the role of search portals and content aggregator of multimedia content is growing. Digital rights management or watermarking technologies may increasingly be used to assure that content is not accessed illegitimately.

Different UCC types (*e.g.* blogs versus video content) have different albeit very similar approaches to monetising UCC. These models can be paired with approaches that remunerate the creators of content (discussed below in the section on economic impacts). Whereas the interest in monetising UCC is growing, most models are still in flux and few providers generate substantial revenues or profits.

There are essentially five approaches to monetise UCC; combinations of these approaches are illustrated in the three cases in Table 5.2 (see also VTT Technical Research Centre of Finland, 2007).

Voluntary donations

In the voluntary donation model, the user makes the content freely available, like that of a musician performing on the street, but solicits donations from users. Such models are currently in place on many sites with a "donate" button, often encouraging those accessing the content to donate to the creator or the institutions (usually online by credit card or via PayPal). A significant number of blogs, wikis, online video and online music creators ask for donations from their audience for activities such as web hosting and site maintenance, or for the content. A common feature of certain non-commercial UCC sites is that they run their operations with quite limited funding (often only the time invested by volunteers and users). Wikipedia, for instance, spent less than USD 750 000 in 2005 to sustain its growth and it frequently draws on donations to finance these costs (beyond the donation of time and expertise made by user contributors).[36] Blogging and citizen journalism sites such as Global Voices Online are supported by bloggers who commit their time, but operating expenses are funded by grants from foundations or even news companies (such as Reuters in the case of Global Voices Online). Such donations of time or money have been the cornerstone of Internet developments in the open source movement (*e.g.* for the support of free Internet browsers) or other user-driven innovations. Voluntary payment models for the promotion of UCC content and platforms based on reciprocity, peer-based reputation and recommendations have been proposed (Regner *et al.*, 2006).

Charging viewers for services

Sites may charge those viewing UCC, whereas the posting of content is free. This can take the form of pay-per-item or subscription models. Popularity has to be high as competing sites are free and making small online payments and entering credit card information may be burdensome or impracticable.

Pay-per-item model. In this model users make per-item (micro) payments to UCC platforms or to the creators themselves to access individual pieces of content. iStockphot, for instance, offers photographs, illustrations and stock video from its user-generated stock for USD 5 each. Platforms exclusively hosting UCC or established digital content sale points (such as online music stores, video-on-demand platforms, or online retailers), for instance, could offer UCC as part of their repertoire on pay-per-item terms. The fact that no shelf space is needed to stock a variety of content facilitates this model.

Subscription model. In this model consumers subscribe to services offering UCC. Paying a subscription to access others' content is rarely used; rather users pay a subscription for enhanced hosting and services for one's own content and access to other's content. In two-tiered subscription services, whereby a user may opt for a "basic" account free of charge that provides a set amount of services or for a "pro" account that users pay a subscription or other fee for. The "pro" accounts provide enhanced features, more (or even unlimited) hosting space, and other options that are attractive to the user.[37] A new approach involves a hosting-based model with a cooperative element, such as Lulu.tv. Users pay for the service provided by the site, but are also remunerated on the basis of the popularity of their content.

Bundling of UCC into existing subscriptions and associated payments may be a more viable option. Cable TV operators, Internet Service Providers (ISPs), digital radio services and other media outlets derive most of their revenue through monthly subscription fees (*e.g.* EUR 29.99 per month for an Internet triple play offer in France). Such operators could opt to carry UCC, either by creating special channels exclusively devoted to UCC (such as the case with FreeTV in France) or by airing a selection of UCC on the regular programs. In both cases, users pay for the UCC content via their usual subscription.

Table 5.2. Three business models: blogs, photos and video

Citizen journalism: AgoraVox (France)	AgoraVox is a European site supporting "citizen journalism" which is currently based on voluntary in-kind contributions. Users submit information and news articles on a voluntary basis. The submitted content is moderated by the small AgoraVox staff and volunteers. Readers also feedback on the reliability of the information. Despite its low-cost model, AgoraVox aims to generate revenues through online-advertising in the near future. Similar citizen journalism sites such as OhmyNews in Korea remunerate their writers. OhmyNews redistributes advertising revenues to writers for very good articles. On OhmyNews readers also directly remunerate citizen journalists through a micro-payment system.
Photo: Flickr (US)	Flickr is funded from advertising and subscriptions. A free account provides the possibility to host a certain number of photos. Advertising is displayed while searching or viewing photos. This revenue is not being shared with users. A "pro" account for USD 24.95 can be subscribed to offering unlimited storage, upload, bandwidth, permanent archiving and an ad-free service. As Flickr is part of Yahoo! it also enhances membership and traffic to other Yahoo! sites. Similar photo sites such as KodakGallery are owned by firms in the photography business. Users can create free accounts. Revenues are generated through the sales of value-added photo services (*e.g.* purchase of prints).
Video: MyVideo (Germany)	The online video sharing site MyVideo derives its revenues mostly from advertising and from licensing its content to third parties. Recently, ProSiebenSat1 Media, Germany's largest commercial TV company, bought a 30% stake in MyVideo. The objective is to secure a share of Internet advertising, to cross-promote content (UCC content on TV, and TV content on UCC platforms) and to identify interesting content for traditional media publishing (*e.g.* talent search show). Video sites such as YouTube have also started licensing content to telecommunication service providers.

Source: OECD based on company information and press reports.

Advertising-based models ("monetising the audience via advertising")

Advertising is often seen as a more likely source of revenue for UCC and a significant driver for UCC. Models based on advertising enable users and hosts to preserve access to the content that is free of charge while bringing in revenue. The economics of such a service are often compared to free web mail where users get a free service, and owners of the service serve advertising to this audience. Payment for the advertising depends on numerous factors: number of users on UCC sites, related website usage (dwell time on site, depth of visit / page views per session / share of repeat visits), or clicks on the actual advertisement banner leading the user to the webpage of the brand being advertised. Viable sustainable business models are only likely to work with a large enough user base to attract enough advertisers and actions by users generating revenue flows for the site.

Services that host UCC make use of advertising on the site (including banners, embedded video ads and branded channels or pages) to generate revenue. The advertisements can be for specific audiences attracted by certain UCC platforms (often popular, young target groups) or linked with certain content being watched by the user. When users search or watch a particular video, related advertising is shown on the side bar, *i.e.* banners or short trailers start as the computer cursor moves across a banner.

Many UCC platforms such as Fanfiction.Net rely on services to drive advertising revenues (*e.g.* Google AdSense, Microsoft, or the service provided by the UCC hosting site itself such as FeedBurner Ad Network for blogs). Google AdSense automatically delivers text and image ads that are targeted to the UCC site, the requested UCC content, the user's geographic location and other factors (for example, travel ads for China when searching for the keyword "China" on a video site).[38] When users click on the ad, that advertisement service receives per-click revenues from the company being advertised. In turn, it then pays the UCC site hosting its ads. Some UCC sites are also redistributing part of this advertising money among those creating or owning the content. These models provide independent UCC sites (some owned by individuals) with access to a large base of advertisers.

Advertising may also be placed within the content, such as within a video. Popular video podcasts also incorporate advertisements where users can click to sites from within the video. Increasingly, "branded channels" have been launched on UCC platforms where users can view content from a special brand or media publisher. Virtual worlds like Calypso allow firms to create and display advertisements.[39]

It is expected that sophisticated targeting techniques will increasingly enable advertisers to create targeted ad messages. The quality of the targeted nature of the advertisement will depend on how well videos or UCC is paired with relevant advertisements. Currently, advertisements are often displayed on the basis of tags and keywords which uploaders create. These may be more or less reliable with some users not creating keywords or using misleading ones to attract more traffic.

UCC platforms have received substantial sums from businesses wanting to advertise to their community. In August 2006, Google agreed to deliver at least USD 900 million in advertising revenue over three and a half years to News Corporation for the right to broker advertising on MySpace and some other sites (van Duyn and Waters, 2006) and Microsoft has also agreed to be the exclusive provider of advertising to Facebook (Sandoval, 2006a).

Although most of the hopes to monetise UCC are currently placed on purely advertising-related business models, it will take time to see whether these models will work (see also VTT Technical Research Centre of Finland, 2007 which argues that social media cannot fully flourish on ad-based models). Advertisers are concerned that the user audience may be accustomed to free content and will migrate to ad-free sites, and some are also concerned with being associated with UCC they cannot control or foresee (*e.g.* a car advertisement being shown before a UCC video about a car accident).

Licensing of content and technology to third parties

Increasingly UCC is being considered for other platforms and licensing content to third parties (*e.g.* TV stations) may be a source of revenue. According to most terms of services of UCC sites, users agree that they have given the site a licence to use the content without payment, sometimes reserving the right to commercially exploit the work (see also below).[40] Sometimes this may include the right of the UCC site to licence the content to third parties but a revenue sharing model between content creators and the UCC site may apply. Increasingly deals to licence content to third parties or to cooperate with third parties to share the content involve mobile carriers (*e.g.* the Verizon and YouTube "Watch on Mobile" service).

Finally, UCC platforms can enter into commercial agreements with third parties to provide their technology to the latter (*e.g.* DailyMotion entering a commercial agreement with the French ISP Neuf Telecom to provide its video sharing service technology). Some UCC platforms (*e.g.* On2 Flix) are back-end service providers to facilitate the process of UCC video services of third parties.

Selling goods and services to community ("monetising the audience via online sales")

Another option is to use the large, captive user base to market own or third party products. Due to network effects, successful UCC sites are likely to have a large user base, and this large audience can be monetised with UCC sites selling items or services directly to their users. Blogging, photo sharing and other sites may sell particular one-off or continued services to their users similarly to the examples in the pay-per-item or the subscription section. UCC platforms such as virtual worlds or social networking services can sell the use of online games, avatars, virtual accessories or even virtual land to their users. The Korean social networking site CyWorld, for instance, receives considerable revenues from the sale of digital items such as decorations for a user profile or furniture for one's virtual "miniroom".[41] Users use Acorns as currency in the CyWorld Shop purchased via credit card.

UCC sites can also co-operate with third parties to allow them to sell directly to their users while taking a share of the revenue. For instance, the Mypurchase service of MySpace will provide the interface for creators to sell their music, taking a portion of sales revenues in exchange. The popular Japanese social networking site Mixi has several approaches, one of which is to allow users to rate and review books CDs, DVDs, games, electronics and other items and linking them directly to Amazon Japan with one click to purchase those items ("social commerce") or to listen to music which can later be bought over iTunes.

UCC platforms could also allow for transactions amongst their users while taking a share of the revenue. Depending on the terms of service, other business models may involve the sale of anonymised information about users and their tastes and behaviour to market research and other firms.

Overall, each of these business and revenue models has advantages and disadvantages, and which will be relatively more successful is still being worked out.

Chapter 6

ECONOMIC AND SOCIAL IMPACTS

The production and consumption of user-created content has been driven by technological developments, more active and participative Internet users and associated social and behavioural changes in use, economic forces and commercial opportunities, and institutional and legal change. However, the actual impacts of UCC and a more participative Internet are in their early stages, and longer-term impacts are unclear even if there are promises of greater changes to come. Furthermore social and behavioural changes are interrelated with economic ones and neither can be seen in isolation. This section provides a first analysis of the economic and social impacts of user-created content to provide the context for the discussion of opportunities and challenges for users, business and policy.

Economic impacts

User-created content is becoming an important economic phenomenon with direct impacts on a widening range of economic activities despite its original non-commercial context. The development and spread of UCC has contributed dramatically to the stock of broadband content. Combined with the attention devoted to it by users, this has had disruptive effects on how content is created and consumed and on established industries and activities supplying content, creating economic opportunities and challenges for both new and established market participants.

This section discusses some of the economic implications of UCC and how the business models outlined in the previous section and efforts to monetise UCC are developing. However, it must be borne in mind that monetisation of UCC is in its infancy, data on business impacts are generally unavailable, and aggregate economic impacts on growth and employment difficult or impossible to quantify.

Table 6.1. Economic incentives and benefits for different UCC value chain participants

Consumer electronics and ICT goods	Selling hardware with new functionality and interoperability for users to create and access content.
Software producers	Providing ICT services and software for creation, hosting and delivery of UCC.
ISPs and web portals	ISPs using UCC to attract customers and build a user base for premium Internet services. Web portals aiming to attract traffic, build Internet audiences and advertising revenues (and avoid losing traffic to UCC-related sites).
UCC platforms and sites	Attracting traffic, building Internet audiences and subscription and advertising revenues; increasing attractiveness for acquisition by third parties.
Creators and users	Facing either non-commercial incentives (to entertain or inform other users, for recognition or fame) or commercial incentives to generate revenue through donations, sale of content or sharing revenue from advertisement-based models. Other users benefitting from free access to content which is entertaining, educational, informational or expands choices (e.g. on purchasing decisions, or advice on various topics).
Traditional media	Participating in UCC online revenues (essentially through advertising-based business models), reaching out to UCC audiences to promote own content. Broadcasting or hosting UCC to retain audience and advertisers. Preventing UCC from decreasing revenues from other content and/or preventing disintermediation.
Professional content creators	Reinventing business models to compete with free web content (e.g. photographs, images).
Search engines	Using UCC audiences to drive advertising revenues while improving searchability.
Web services that benefit from UCC	Using UCC to build more attractive websites and customer services and information (e.g. a travel agency or hotel chain that encourages users to post pictures and share appreciations).
Advertising	Participating in increasing online advertising directed at communities on UCC platforms. Using UCC content in advertising campaigns.
Marketing and brands	Expanding customer loyalty through promotion of brands on social networking sites or through advertising to UCC communities.

Different value chain participants have different economic incentives and derive different benefits from the expansion of UCC, as shown in Table 6.1.[42] While press reports have centred largely on the impact of UCC on traditional media firms, more immediate impacts are on non-media firms and users. The more immediate impacts on market entry, growth and employment are initially with ICT goods and services providers and newly formed UCC platforms which have attracted significant start-up venture capital investment and been the target of acquisitions. Other firms active in monetising UCC are search engines, portals and aggregators attempting to develop UCC search, aggregation and distribution business models, often based on online advertising.

Impacts are also being felt by professional (often free-lance) content producers including journalists, writers, photographers, video-producers and others who are competing with freely available web content.

Consumer electronics, ICT hardware, software, network service and platforms

The consumer electronics industry, the telecommunications industry (including Internet service providers and increasingly mobile operators), firms creating the tools and software to edit and publish content and information technology in general have benefited from UCC as users buy digital cameras and other content-creating tools, software, and broadband access to watch and create content. In addition, innovative new firms are entering the market and creating employment with new goods and services.

The ICT sector grew 6% in 2006, with much higher growth in digital storage, new portable consumer products, Internet activities related to the participative web and new digital products (OECD, 2006a). In particular, digital entertainment is currently a high growth market, with a growing global market for digital consumer appliances and a comeback in the last three years for the electronics industry (CEA, 2006).[43] The detailed figures show users spending more on flash and hard-drives, portable MP3 audio players, digital cameras, new mobile equipment, display devices and related accessories despite price declines. New cross-platform application technologies allowing playback of popular web video content on TV, mobile phones, and other devices will drive further growth of ICT products.

The creation, upload and download of ever-larger content files are important drivers for the network services industry and other infrastructure providers. The need for greater speed, wireless and mobile access is increasing network revenues and is going some way towards replacing loss of traditional fixed-line voice revenues (OECD, 2006d, 2007a). To move into more value-added services, telecommunications operators are also increasingly interested in hosting and providing UCC-related services. For example, Vodafone has announced that it will try to unite virtual and real communication, allowing people to talk through their avatars without necessarily sharing their real personal details.

The popularity of video sharing services has led to new types of application service providers ("broadband video application service providers").[44] Industry observers estimated that YouTube was already streaming 40 million videos and 200 terabytes of data per day in early 2006,[45] and new firms such as Brightgrove, Entriq and Maven Networks offered publishing, syndication, commerce, content management, security and other platform components in the form of software-as-a-service. The market was projected to reach USD 1.9 billion in 2011 (ABI Research, 2006). Firms such as Limelight Networks providing content delivery to popular UCC sites report strong growth (USD 14 million revenues in the second quarter of 2006).[46] Firms providing one-stop video upload, converting and transmission of optimised content services while reducing bandwidth requirements will see a large demand.

Similarly, spending is increasing on tools, ICT services and software which enable users to create, edit, locate, post and monetise audio and video files (Apple garage, Grouper, Jumpcut, Ripple Share) and on video compression software. Firms which are producing new tools such as recommendation engines, or which allow the creation of playlists, blogs and podcasting are entering the market, attracting investments and increasing revenues. In Japan the blog-related market was estimated to be about JPY 137.7 billion (USD 1.2 billion) in 2006 (MIC, 2006).[47] This includes tools to publish and host blogs, blog software and related advertising, and mobile phone applications which make it easier to share photos and videos with other users and networking sites (*e.g.* developed by Shozu.com) are growing. The demand for digital rights management, watermarking technologies and other software tools will also increase.

Users/creators

New models are also emerging to remunerate content creators (*i.e.* creators of original works) and, for example, surveys of younger age groups point to greater willingness to place advertisements at the end of videos, to feature brands in the video and to derive revenues from their work.[48] A distinction must be made between models that provide revenue for the creators themselves and models which entail revenue sharing between creator and host. Initially, the possibilities for users to collect revenue directly without intermediaries were limited and a number of co-operative models were developed. A co-operative-based model is one where the creators contribute money to the service and this revenue is redistributed among the creators. A hybrid model, such as Lulu.tv, combines the co-operative method with shared advertising-based revenues; users-creators pay for the service provided by the site, but then they are remunerated on the basis of the popularity of their content.

There is also the possibility of creators being remunerated via flat-fee subscription services. New services have arisen which allow for unlimited uploading and exchange of various content and which are financed by monthly fees from subscribers or Internet Service Providers. The Digital Media Exchange (DMX) is such an example based on P2P technology and operated as a non-profit co-operative.[49] As well as unlimited exchange it allows users to make derivative works from the content while being copyright compliant. DMX collects monthly fees from subscribers or their ISPs, and pays all of the collected fees to content suppliers.[50]

Most revenues for creators are likely to come from revenue-sharing models between creator and host based on advertising revenues and UCC platforms are usually involved in remunerating content creators. In the case of creator-based advertising, users would utilise advertising on, within, or surrounding the content they created. For example, bloggers may put Google AdSense advertisements on their site. Many new companies and software tools are allowing users to post content and be remunerated (*e.g.* Ripple Share). In Korea, Mgoon has introduced a website "tag story" where users post their own content and share advertising revenues with the company. Korean Shotech UCC site has started to share 30-40% of its advertising proceeds with UCC makers. Other startups such as Revver, Feedburner, Blip.tv and Panjea.com share at least 50% of their ad revenue with users who create their content.[51] It is expected that the quality of the UCC can be improved if creators are remunerated.

Finally, professional and paying careers can arise for users creating content. For example, students in popular lip-synching videos were later hired for commercials. Remixers, bloggers or video podcasters have been hired by major music or media companies.

Users who consume the content usually benefit from free access to more diverse content which may be entertaining, educational or serve other purposes. In particular, information and knowledge commons mentioned earlier can add significantly to consumer welfare and entertainment.

Traditional media

The impacts of the shift to Internet-based media – both in terms of challenges and opportunities – are only starting to affect content publishers and broadcasters. Moreover, the involvement of media publishers in the hosting and diffusion of UCC is at an early stage.

At the outset, UCC may have been perceived as a challenge and competition to established media firms. First, UCC sites often host unauthorised content from media publishers without consent and without remunerating the rights-holders (*e.g.* full or part clips direct from TV shows which fall outside of the UCC definition adopted here). Thus traditional media publishers have a legitimate conflict of interest with UCC platforms and potentially face declining revenues due to illegitimate access to their content.

Second, users may create and watch UCC at the expense of other traditional media such as TV. This is a source of concern, given that many traditional media activities are based on attracting large audiences that generate advertising revenues to support the creation of content to attract further audiences and so on. As opposed to traditional media products, UCC can sometimes be produced very cheaply while still drawing a large returning audience. For example, Rocketboom is a popular three minute, daily video blog distributed via RSS and produced with a video camera, a laptop, some accessories, but with no additional overhead, promotion or large bandwidth costs.

Third, traditional media sources may be used more selectively. Most of the existing studies demonstrate that time spent on the Internet reduces the consumption of offline media (see OECD, 2004a). Figure 6.1, for example, shows that Internet use in the United Kingdom has led to a reduction of attention devoted to TV, newspapers and the radio. This effect is particularly notable in the 15-24 age bracket, which make considerably less use of traditional media and their viewing has declined more, with newspapers, magazines and radio particularly affected by this change in media consumption.

However, these surveys are of the Internet broadly as a potential substitute for established media not specifically UCC. In addition, media such as the major television channels face increasing competition from an ever greater number of new or specialist television channels, newspapers, and from other new media (*e.g.* online games, digital radio). Consequently, some of the concerns directed toward UCC may be a reaction to a new distribution medium and changing user habits which are general and hard to predict. Media firms that are used to "broadcasting" the same content to large audiences (one-to-many) are adapting to users who may be in search of more targeted, interactive or personalised on-demand "narrowcast" content (one-to-one). Content publishers will increasingly compete and co-operate with many entertainment forms (including UCC) for more fragmented audiences and advertising revenues.

Figure 6.1. Reduced consumption of offline media in the UK driven by Internet use, April 2006, % of respondents

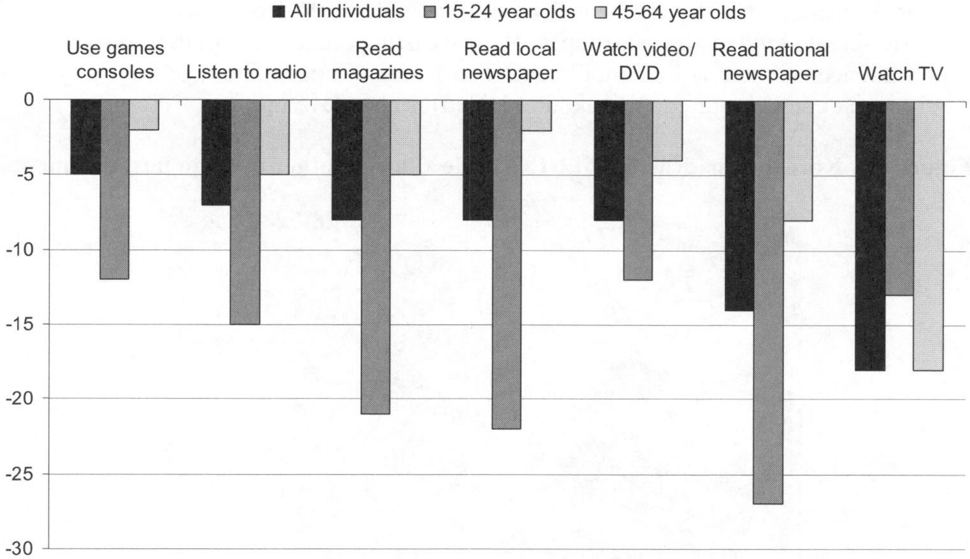

Question: Since using the Internet for the first time, which if any of the following activities do you believe you undertake less?
Source: Ofcom research, April 2006.

Overall, UCC has already begun to have an impact on traditional media industries and vice versa, and is increasingly seen as a market reality. Progressively, the advent of UCC marks a shift among media organisations from creating on-line content to creating the facilities and framework for non-media professionals to publish in more prominent places or at least to take existing UCC into account when publishing their own products. For a start, traditional media have made their own websites and services more interactive, enabling users to comment, give feedback, rate particular articles and content and diffuse their content via participative web technologies such as RSS feeds, podcasts or host blogs on their sites.

Television companies increasingly select UCC for use on their usual channels or web pages; a trend facilitated by technology which allows viewing broadband content on television. For example, in France TF1 has launched the WAT site which allows users to upload content for later potential TV diffusion against remuneration (see also M6 Web in France). CBS Television in the US is co-operating with YouTube to select and air the best viewer-videos. There is also an increasing number of TV channels being created to show only UCC (*e.g.* Current TV), and there are examples of Korean terrestrial broadcasters using UCC while also allowing users to post comments (see Figure 6.2). Sony will introduce TV receivers which can access Internet video content (*i.e.* Bravia Internet Video Link) including user-created videos from Grouper. Narrowcasting content to target a specific audience or topic (Election TV and The People Choose 2006, or specialty interest content aggregators like Pet Video) is also developing.

Figure 6.2. Korean broadcasting of UCC while allowing other users to post comments

Source: www.afreeca.com.

Major media companies also have acquired or are in the process of acquiring content from or licensing content to UCC platforms to generate revenues and explore ways of extending their on-air programmes and brands to UCC platforms. In some recent deals, considerable amounts have been paid by video-sharing platforms to media firms to continue to host their content. Under these agreements, video sharing sites are offering free and full access to music videos (*e.g.* Vivendi Universal) or television content (from CBS, NBC Universal) – sometimes sharing related advertising revenue with the content owners, sometimes only relying on the promotional effects. Increasingly, traditional media firms have opted to advertise their content within UCC sites by giving access to trailers, free samples of some content, etc.

The impact of the increased involvement of traditional media and established Internet firms in UCC sites is unclear. On the one hand, they provide the necessary "backbone" to UCC platforms and potential remuneration to creators. On the other hand, users may migrate away from UCC platforms which have too much centralised control, too many intrusive advertisements and forms of branding, or which develop into video delivery platforms for commercial content producers crowding out UCC.

Newspapers and other print media (which will be discussed in greater detail in a separate OECD Study on Online News Distribution) have also started to change their way of reporting or commenting on the news and – in some cases – to aggregate UCC, *e.g.* with discussion fora, interactive blogs and other interactive features ("citizen journalists").[52]

Professional content creators

The impact of UCC on independent or syndicated content producers who previously did not face competition from Internet users or non-professional creators has been increasingly discussed on blogs and other fora. Photographers (including independent agencies), graphic designers, free-lance journalists and other professionals who provide pictures, news videos, articles or other content face competition from freely provided or low-cost UCC or other amateur-created content. In terms of photos and images, for example, the web offers a vast amount of copyright-free content (see Flickr, iSTockphot or how user-created blogs and videos supplant some editorial content in newspapers or online media ventures) which has decreased the average unit cost of images and competes with commercial image providers. In terms of video footage and news articles, citizen journalists are in some ways also competing directly with freelance and other journalists for the reader's attention.

Overall, it is too early to be able to quantify the economic impacts on professional content producers. On the one hand, the suppliers of such professional content will have to readjust, some no longer able to make a living from their original work. In the case of agencies for photography and images, these seem to be increasingly concentrated into a few providers (*e.g.* Getty Images, Corbis, Jupiter Images), although whether this is a natural evolution due to high classification, archiving and retrieval costs and related economies of scale, or partly due to the increasing availability of amateur content is unclear. On the other hand, cheaper or free access to such content also reduces costs for business and consumers.

Search engines and advertising

Search engines and advertisers see the potential in capitalising on the user base of UCC sites. Worldwide growth of online ads was for example expected to be around 35% in 2006 to reach USD 11.6 billion (van Duyn, 2006, see also forthcoming OECD Online Advertising Study). While still small, it is expected that advertisement-supported UCC will be one of the main drivers of website revenue (eMarketer, 2006a, 2006b). This trend has been supported by the recent acquisition of YouTube by Google, and SK Telecom which also owns the popular CyWorld is also planning on developing a search engine that combines search and UCC.[53]

Services that capitalise on UCC

Commercial websites and services are emerging that allow users to contribute their content, increasing the overall interactivity and customer value of the site. Increasingly travel services (*e.g.* hotel websites, flight booking sites), real estate services and other websites allow users to share their experiences (*i.e.* "word of mouth"), pictures and ratings.

Marketing and brands

On social networking sites, brands increasingly create special sub-sites and areas with social branding. Members of these sites may join brand groups and display logos on their personal profiles. Finally, advertisements can take a different form in virtual worlds. Second Life, for instance, has news companies such as *Wired magazine*, CNET, Reuters, and firms such as Adidas and Toyota buy virtual land (in this case advertising space), to hold press conferences and to advertise their brands (Figure 6.3). Another form of branding and advertising takes place when artists use UCC sites to build prominence and attract the attention of record labels, film studios, newspapers and the like.

Figure 6.3. Reuters in Second Life

Source: Second Life.

Established brands have also begun experimenting with new ways to integrate UCC into their advertisements. For example, when a video portraying the explosive combination of Mentos and Diet Coke spread on the video-sharing site Revver, Mentos responded by placing advertising within the video. Initiatives by a US car manufacturer and by a beauty product supplier[54] have encouraged users to create their own commercials. Other firms have created sites on which users can create humorous content.[55] There is, however, no way to know what users will do on such sites, with brands potentially losing control of their image and being subject to greater information exchange, for example between users intent on revealing flaws in products and bad service sometimes justified but sometimes not.

Market research is also using UCC to gather and use the information exchanged between users on the Internet – for example on brand reputation, customer service and product performance – to improve customer service, marketing campaigns, brand image and the products themselves. For example, Nielsen BuzzMetrics provides technology to help companies understand and use increasing consumer activity and content on the Internet.

Use of UCC and participative web tools in business

While this study focuses on UCC, tools associated with the participative web are increasingly used by businesses and professions in their wider economic activities. Blogs, wikis, podcasting, tagging technologies, and techniques from community and social networking sites can be important tools to improve the efficiency of knowledge worker collaboration, advance product development and raise the quality of interactions with goods and service users and consumers. As the use of these tools expands, significant new uses of and markets for participative web technologies will develop and there are likely to be potential impacts on business organisation, behaviour and productivity.

Social impacts

Despite growing economic impacts of UCC, the creation of content by users was initially perceived as a mainly social phenomenon with mainly social implications, raising a host of questions, opportunities and challenges.

On the side of opportunities, changes in the way users produce, distribute, access and use information, knowledge, culture and entertainment results in three cross-cutting trends: increased user autonomy; increased participation, and increased diversity, with structural impacts on the cultural, social and political spheres. The Internet and UCC may also change the nature of communication ("who communicates with whom under what conditions and whose discretion"; Benkler, 2006) and related social relation-ships. Challenges cover issues of inclusion, cultural fragmentation and issues related to security, privacy and content quality.

Increased user autonomy, participation and communication

The rise of UCC provides new ways for how information, knowledge and culture is developed and exchanged, potentially at lower cost (MIC, 2006; Benkler, 2006). The Internet has altered the nature and the economics of information production as entry barriers for content creation have significantly declined or vanished and led to the democratisation of media production (sometimes referred to as the "rise or return of amateurs"), distribution costs have declined dramatically, user costs are lower, and there is much greater diversity of works with shelf space in the digital media being almost limitless.

These changes imply a shift away from simple passive consumption of broadcasting and other mass distribution media models ("couch potatoes") to more active choosing, interacting and creating content (Lessig, 2004) and a shift to a participatory "culture". Technological change empowers individuals to "tell their stories", to produce cultural goods such as music and video and to transform the information and media content environment surrounding them (Benkler, 2006, OECD, 2006a, 2006b, Fisher, 2004, 2006). Users may derive a higher value from this content consumption as it may be more personalised and on-demand (one-to-one, "narrowcast") with users having greater control over it.

Furthermore, the changed structure of communication and resulting active relationships built around exchange are argued to have important impacts on how citizens and users communicate and express themselves as well as possible positive impacts on social ties and social structures.

Cultural impacts

The cultural impacts of this social phenomenon seem far-reaching with culture at different levels being affected by new and different ways of creating and diffusing content and new interactions developing between creators, users and consumers (OECD, 2006a, 2006b). According to the long tail effect, a more diverse and substantially increased set of cultural content will find niche audiences and users, potentially expanding creativity. Some artists have already become celebrities while using UCC platforms to gain recognition.[56] This phenomenon applies to all genres. In Korea, for instance, one portal alone had over 150 000 literature-related forums where classic and novel genres were created and commented on by amateur critics (National Internet Development Agency of Korea, 2006). Talent agencies, media publishers and Internet sites now use UCC platforms to discover new talent (screenwriters, directors, musicians).[57] Even if the platforms lead to recognition and development of only a few outstanding content creators, this may still be considered a cultural enrichment.

The availability and diversity of (local) content in diverse languages is increasing, with dynamic user bases in most OECD and many non-OECD countries. As more participate in building and cultivating culture and the democracy associated with egalitarian cultural development, greater general identification of users with culture and society may develop and less alienation result (Benkler, 2006; Fisher, 2006).

Citizenship engagement and politics

The Internet can be an open platform enriching the diversity of opinions, political and societal debate, free flow of information and freedom of expression. UCC is in many ways a form of personal expression and free speech, and as such, it may be used for critical, political and social ends. It has also been argued that the "democratisation of access to media outlets" fulfils an increasingly important role for democracy, individual freedom, political discourse, and justice (Balkin, 2004; Fisher, 2004; Lessig, 2004; Benkler, 2006). As users raise questions and enquire, and as new decentralised approaches to content creation are adopted, the political debate, transparency and also certain "watchdog" functions may be enhanced (*c.f.* websites like Meetup and Pledgebank which facilitate collective action on political and social issues by civil society).

Citizen journalism, for instance, allows users to influence or create news, potentially on similar terms as newspapers, companies or other major entities. Creators of UCC have brought attention to issues that may have otherwise gone un-noticed (*e.g.* the online circulation of video files of politicians making racist remarks). Bloggers and other users on sites such as AgoraVox (see Figure 6.4) have taken on the role of grassroots reporters and fact-checkers which influences the content in traditional media (Gill, 2005). Effects may include a greater call for accuracy within the mainstream media, as users point out inaccuracies and flaws online. UCC may also provide a way to gain the attention when none previously existed (for example, protest movies against particular events). In some cases issues are covered in great detail which would not be otherwise (*e.g.* a blog that specialises in human rights issues in country x or media reporting on alleged wrongdoings of influential persons or companies). GlobalVoices.com, for example, aims to redress insufficiencies in media attention by using weblogs, wikis, podcasts, tags, aggregators and online chats to call attention to conversations and points of views from non-English speaking communities. Often when unexpected events occur, the only source of immediate documentation may be users with their mobile phone cameras.

Figure 6.4. AgoraVox

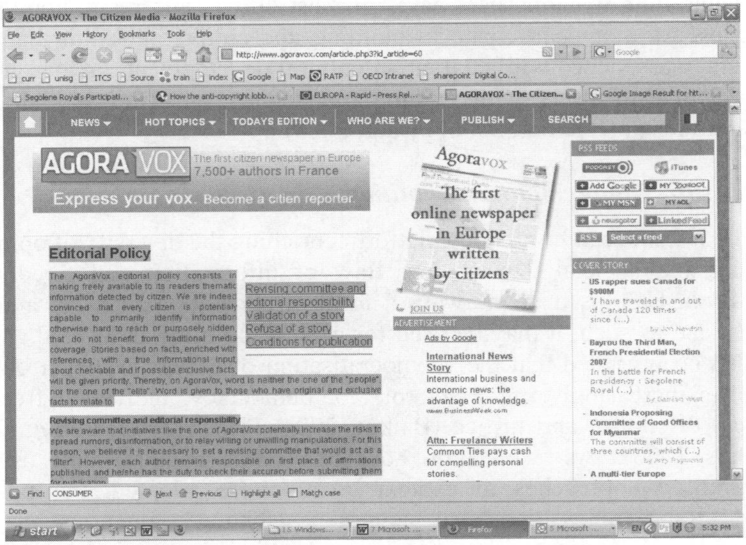

Source: www.agoravox.com.

UCC has been seen to be important in politics and has not gone unnoticed by politicians (see Box 6.1). On the one hand, blogs, social networking sites, and virtual worlds can be platforms for exchanging political views, provoking debate and sharing knowledge about societal questions. On the other hand, they can also be very directly involved in the partisan political process itself. Recently, popular social networks have covered political campaigns, urged young users to vote and have staged related debates. In the US these platforms have been active in getting youth to register to vote or using video contests to invite thoughts on national policy (e.g. the "MyState of the Union" initiative by MySpace). Social networking and other UCC sites are increasingly recognised tools to engage the electorate as evidenced by the discussions but also by the increased presence of politicians on sites hosting UCC. In Korea, as of January 2006, 60% of all lawmakers had blogged (National Internet Development Agency of Korea, 2006). The medium- to long-term impact of the participative web on political systems merits further study.

Box 6.1. Politics, news and blogging in France

In France, blogging has become a staple for politicians, and the leading contenders in the 2007 presidential election, including Segolène Royal[58] and Nicolas Sarkozy[59] maintained online journals. Nicolas Sarkozy's appearance in a video podcast in December 2005,[60] the first by a leading politician in France, was seen by many as a major step in embracing new forms of media. Socialist candidate Segolène Royal made her website, blogs, video podcasts and even virtual town hall meetings in a building on Second Life a part of her campaign of "participative democracy". Other prominent uses include blogging for activist and political ends, such as critiquing gender roles in advertising, debating the rejection of the European Constitution, or commenting on the country's labour laws. French newspapers such as Le Monde, have set up a blogging service where users maintain online journals on the newspaper's website.

Source: MediaMetrie and various press reports (including Crampton, 2006 and Matlack, 2005a,b).

Educational and information impact

Multiple users can very effectively develop information and knowledge resources, often for free, that are potentially important ways to provide citizens, students and consumers with more useful and more detailed information. User-created educational content tends to be developed collaboratively and to encourage sharing and peer-production of ideas, opinions, information and knowledge. Provided that it is accurate, the availability of large amounts of freely accessible information such as Wikipedia, Creative Commons materials, freely available pictures on photo-sharing sites, and websites created by individuals (*e.g.* a former teacher providing a website on American history, a naval officer authoring podcasts on different ships) can have positive educational impacts. Discussion fora on various topics as well as product reviews increase the general level of knowledge and potentially lead to more informed user and consumer decisions (*e.g.* Amazon's book recommendations).

Educational projects can also build on participative web technologies such as podcasts to improve the quality and reach of education. Schools and universities may make use of wikis for assignments, group projects, posting notes and syllabi, and creating online resources. The Harvard Extension School and other universities and groups of students, for example, are using UCC platforms such as virtual worlds to facilitate projects (Figure 6.5).[61] Students and educators including in developing countries access resources and knowledge mostly free of charge.

Figure 6.5. Harvard Law virtual distance education classroom

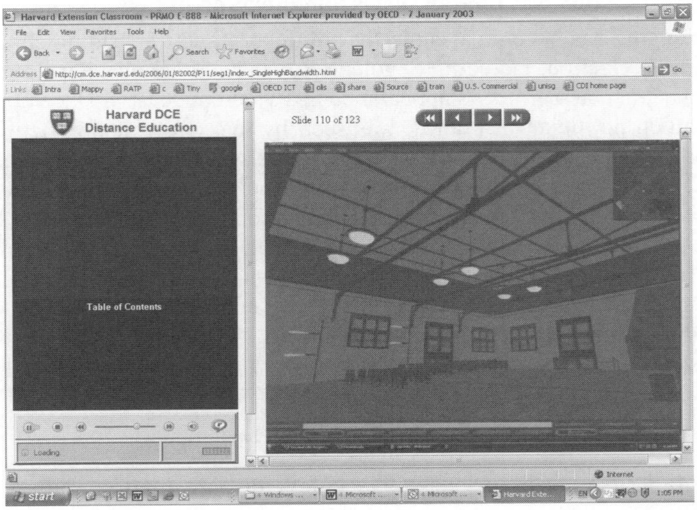

Note: See also Second Life's educational platform at http://secondlife.com/education.

While accuracy and quality of information are well-assured in established institutional contexts, they may be a problem in Internet-based settings where everybody can contribute without detailed checks and balances (see later sections). Wrong information or mistakes (such as, for example, incorrect historical dates) can be widely diffused by the Internet, and discussion fora and product reviews can spread misleading or false information.

Impact on ICT and other skills

The creation and the use of UCC are likely to improve ICT skills, especially in younger age-groups who will need these advanced skills later. Furthermore, the actual process of creating content and viewing it, may create specialised skills such as video filming and editing, acting, writing (*e.g.* novels, poems). Participation in blogs, citizen journalism, critical videos concerning public events or politics and confrontation of different opinions may arouse critical minds and interest in debate.

Social and legal challenges

The Internet and UCC may also raise social challenges. A greater gap between digitally literate users and others (elderly, handicapped, poor people) may occur, accentuating social fragmentation and intergenerational gaps. Cultural fragmentation may also be accentuated with UCC encouraging greater individualisation of the cultural environment with citizens decreasingly sharing a common core of cultural values, exacerbating the effects of the multiplication of media channels and the diminishing role of a few national broadcasting channels for political discourse and shared national values.[62]

Other challenges relate to information accuracy and content quality (including the problem of inappropriate or illegal content), privacy issues, safety on the Internet and possibly adverse impacts of intensive Internet use. In terms of legal challenges, the nature of the Internet and the ease with which content on the Internet can be reproduced and distributed makes online UCC, and all online content, particularly susceptible to copyright infringement. These challenges are discussed in the next chapter.

Chapter 7

OPPORTUNITIES AND CHALLENGES FOR USERS, BUSINESS AND POLICY

User-created content is growing rapidly, and as broadband penetration spreads and ICT user skills develop it will diffuse through all age groups and activities. However the rise of the participative web in general and UCC in particular raise opportunities and challenges for users, businesses and governments which are the focus of this last section. A common set of digital content policy areas has been developed based on detailed analysis of digital content sectors and activities (see Box 7.1) and these policy areas are used as a framework for analysis of various aspects of UCC creation, distribution and access. There are also issues which were not raised in other digital content sectors, including: How sustainable is the UCC phenomenon? Are there bottlenecks to its creation and diffusion? What problems are raised by UCC? And finally should these impediments be removed? If so, how and by who?

This section explores these issues and challenges. The starting point is the potential of UCC and the question of how to foster economic, social, and cultural benefits while defining the boundaries of legitimate use and creating a safe Internet environment. [63] The analysis includes a review of the terms of service of 15 commonly used English-language commercial UCC platforms to provide indicative insights into the legal and policy setting of the most popular ways UCC is hosted, although users can also use non-commercial platforms or host their content themselves and in these cases the terms of services analysed here do not necessarily apply.[64]

Box 7.1. Digital content policies

1. Enhancing R&D, innovation and technology in content, networks, software and new technologies.

2. Developing a competitive, non-discriminatory framework environment (*i.e.* value chain and business model issues).

3. Enhancing the infrastructure (*e.g.* technology for digital content delivery, standards and interoperability).

4. Business and regulatory environments that balance the interests of suppliers and users, in areas such as the protection of intellectual property rights and digital rights management, without disadvantaging innovative e-business models.

5. Governments as producers and users of content (*e.g.* commercial re-use of public sector information).

6. Conceptualisation, classification and measurement issues.

Source: OECD (2006c), "Digital Broadband Content Strategies and Policies", www.oecd.org/dataoecd/54/36/36854975.pdf.

Enhancing R&D, innovation and technology

R&D and innovation

The creation of user-created content relies heavily on the availability of various new participative web technologies, for example tagging, group rating, content distribution (content packaging and management; digital asset management), interactive web applications and content management systems, 3D graphic production and digital animation (see earlier section on drivers of UCC), and innovations in networks and consumer electronics. Often technologies will be the solution to many business and policy challenges such as combating spam, ensuring a safe Internet experience (implementing parental controls, age ratings), and securing intellectual property rights. These technologies for content creation and diffusion are increasingly R&D-intensive (faster networks, new platforms, content-intensive products, data-base management) and the challenge is to establish business settings, policies and approaches that encourage innovation.

Governments have an on-going role encouraging R&D and innovation. Market participants and institutions create and commercialise innovative products, but governments support basic R&D, address market failures and provide an environment conducive to R&D and innovation (*e.g.* R&D tax incentives, specific R&D support, see OECD, 2006a) and very often encourage linkages between commercial and not-for-profit R&D and innovation-related activities.[65] Basic R&D receives major public sector support, *e.g.* for energy, environment, health, military and space, and much of this involves new components, devices, networks, imaging, software for virtual environments, 3-D modelling, etc. which impact UCC in various ways. To the extent that there are general market failures in the development of new software or platforms, governments may aim to foster an innovative business environment without choosing particular commercial or non-commercial technologies.

Ensuring technological and other spillovers

Content creation and delivery technologies and user-created content itself are increasingly used in other, non-entertainment, sectors. There may be considerable technological spillovers from UCC technologies and approachesdeveloped in areas such as games imaging and virtual world technology to medical and other fields.

Creative environments, skills, education and training

The creation of UCC and the necessary services and technology to support it rest on creative environments, skills and education. A central question is if and how governments can encourage and promote UCC-related specific or general skills, and if there are new models to foster and provide incentives to creativity. Governments have a major role in developing skills via secondary and tertiary education and vocational training. For the creation of UCC, basic and sometimes more advanced ICT skills are needed, and advanced skills are necessary to develop the surrounding technologies. Furthermore, younger generations will more readily have the required middle-level ICT skills, but older generations, the disabled or poorer groups may warrant special efforts.

Fostering local and diverse content

Cultural and language issues are seen as important in the development of digital content, particularly for smaller countries and linguistic and cultural minorities. Many OECD countries have rapidly developing markets for UCC and related services, and the development of user-created content may increase availability of specialised local and minority content in diverse languages. With lower access barriers and increased demand for content downstream and lower entry barriers in supply upstream, the creation of cultural content and identification of new creators could potentially be enhanced.[66]

There is significant government support for local content development where market failures are perceived to exist, and public institutions may have a role in driving the creation of UCC. In some countries public broadcasters have initiatives allowing citizens to download their content, and rip, mix, and burn it (see in the UK the example of the British Broadcasting Corporation, BBC).[67] Cultural policies and institutions such as museums, musical conservatoria, and other cultural and educational institutions may also encourage innovation around UCC with the public policy objective of fostering creativity and cultural expression. Support programmes for the creation and diffusion of local content may need to further take account of the potential behind UCC and its associated diffusion and use.

On the other hand, in the United Kingdom, making available free content for downloading, remixing and other uses has triggered concerns that freely provided public content could "crowd out" the creation of commercial content with free public content competing on an unequal footing with commercial content.

Developing competitive, non-discriminatory policy frameworks

The development of user-created content has thrived on wide-spread access to networks, software, content and other services, and commercial businesses are increasingly involved in supporting the creation, search, aggregation, filtering, hosting and diffusion of UCC. To encourage further experimentation and competition in value chains and business models, it is essential to maintain and further develop competitive, non-discriminatory policy frameworks and a pro-innovation business environment. This begins with the market for telecommunication services but extends to other business activities (*e.g.* traditional and new content industry entities, Internet portals, search engines).

Control over parts of the value chain should not unduly restrict new entrants or users creating content, in particular small firms. This holds particularly true in new fields such as digital rights management, an increasing concentration in search services, and technologies and services which limit or prevent interoperability. Very strong network effects, potential for lock-in and high switching costs have to be taken into account when making competition-related assessments of UCC services which have a critical mass of users.

However, new forms of digital content innovations are often based less on traditional scale advantages and large initial capital investments and more on decentralised creativity, organisational innovation and new business models for content production and diffusion (OECD, 2006a, 2006b). These factors favour new entrants, particularly for new platform aggregation models, where content owners had no legacy advantages (IBM, 2007). Very popular services were started by a small group of individuals and rapidly competed with established entities.

Maintaining the open and collaborative nature of the Internet is a necessary condition for the further evolution of UCC. The question is whether the Internet will preserve its open nature with interoperable services or whether it may evolve into "walled gardens" which may be preferred by some users for reasons of simplicity, quality and security, and the role of policies is to ensure that the users can choose between these different options. Finally, the growth and development of UCC may have to be taken into account when determining policy on the prioritisation of network traffic. It is unlikely that individual users creating content on an informal basis would have the ability or funds to negotiate agreements with ISPs.

Enhancing the infrastructure

Broadband access

Widespread affordable broadband availability and access is a general policy target.[68] Broadband policies to ensure (regional) coverage and equal access to infrastructure and applications on fair terms and at competitive prices to all communities, irrespective of location, are general policy aims, as is a regulatory environment which encourages investment and competition in communication networks and technologies and which is adapted to next generation networks.[69] Initiatives such as the provision of municipal broadband networks can, in the absence of market solutions providing similar access, also be beneficial to the creation of UCC (including the accessibility of broadband services to the disabled).

One key technical challenge for the evolution of UCC is the low consumer availability of symmetrical networks. The majority of Internet connections are Asymmetric Digital Subscriber Line (ADSL) and cable services which are "asymmetric", with the volume of data flow greater in one direction than the other. Providers usually market services for consumers to connect to the Internet in a relatively passive mode, with the higher speed for download from the Internet without running servers for high speed in the other direction.[70] There is now greater potential for more symmetrical content download and upload, and the current infrastructure is not conducive to more symmetrical user behaviour. The deployment of new distribution technologies such as optical fibre (as in Japan and Korea) can help overcome this problem.

Convergence and regulation

Ensuring effective competition and continued liberalisation in infrastructure, network services and applications in the face of convergence across different technological platforms is a key policy challenge. Many new services have appeared which support the distribution of video material on an on-demand and one-to-one basis. These are often called non-linear services where the user decides upon the moment in time when a specific programme is transmitted on the basis of a choice of content selected by the media service provider. The distinction is between TV broadcast or linear services ("push content"), on one hand and non-linear or on-demand services ("pull content"). ISPs, new video services and others are now involved in the creation, hosting and diffusion of such one-to-one "pull content". This technological and business convergence is putting existing regulatory arrangements and the separation between broadcasting and telecommunications regulations to a test, particularly as telecommunication regulations mainly focus on establishing competition, while broadcasting policy also tries to achieve wider public policy objectives (*e.g.* the protection of minors, cultural diversity) (OECD, 2004b, 2006e, 2007a).

Many OECD countries are in the process of realigning their regulatory regimes to deal with convergence as Internet content has become much more widespread.[71] The essential question, while taking into account the particular nature of on-demand one-to-one video services, is to determine up to what point the new services should be subjected to similar rules as those applicable to traditional broadcasters or rules inspired by those.[72] Examples of such regulations are broadcasting and production quotas (*e.g.* transmission time reserved for works of independent producers), rules on television advertising and sponsorship (*e.g.* maximum advertisement time per daily transmission, identification of advertising, rules on surreptitious advertising, restrictions on certain advertisements such as alcohol and product placement and sponsoring), the protection of minors, rules on incitement to hatred, the right of reply (*i.e.* a person whose legitimate interests have been damaged by an assertion of incorrect facts in a television programme must have a right of reply), and how events of major importance for society have to be treated.

With the emergence of increasingly advertisement-based business models and unsolicited email and marketing messages, rules on advertising will play a particular role in the UCC environment (in particular product placement in virtual worlds, and in the context of advertising to children).

A question which has also been raised is if and how UCC types and platforms can fulfil, extend or complement more effectively certain functions which up to now have been attributed to public broadcasting (public debate, social cohesion etc.).

Regulatory environment

User-created content raises a range of issues with respect to the business and regulatory environment in which it is created. The blurring of differences between content users and producers needs also to be kept in mind. If users increasingly derive non-pecuniary and pecuniary benefits from content creation and they become actual producers / commercial entities, their legal and regulatory treatment will change (*e.g.* in areas of consumer protection, intellectual property rights, taxation, and other producer responsibilities).

Intellectual property rights and user-created content

Copyright law is intended to encourage the creation and dissemination of authors' works of and thereby to promote cultural and economic development. From an economic perspective, copyright is designed to provide exclusive rights for a limited time to authors to recompense their creative effort in return for enabling their works to be widely appreciated and to encourage further creativity. This section discusses the salient intellectual property rights (IPR) issues in the areas of UCC and points to areas where further work may be needed.[73]

For a work to enjoy copyright protection, it must be an original creative expression of the author.[74] Generally, copyrights confer on authors and/or right-holders a set of exclusive rights, *i.e.* the control over reproductions, the preparation of derivative works (*i.e.* adaptations), distribution to the public, public performances and public display. In some countries copyrights are also intended to protect the rights of integrity and attribution sometimes identified as the moral rights of authors (*i.e.* ability of authors to control the eventual fate of their works). These rights expire when the copyright term ends and a work falls into the public domain. Moral rights may continue even after the economic rights have expired (for example, in France).

Copyright regimes in OECD countries aim at balancing a creator's exclusive rights and the public interest in the creation, access to and wide dissemination of knowledge and creative works. This is pursued through exceptions and limitations to the creator's rights. These exceptions and limitations may be specific statutory exceptions and limitations which may or may not include fair use and fair dealing principles. In addition, information in the public domain is not subject to copyright protection. Under certain circumstances, exceptions and limitations allow the reproduction and adaptation of copyrighted works without the authorisation of rights-holders. Both exclusive rights and exceptions and limitations have been clarified to apply to existing norms in the new digital environment, notably through the ratification of the WIPO Internet Treaties[75] (see WIPO,

2003; OECD, 2005b). The Recommendation of the OECD Council on *Broadband Development* recommends that member countries should implement regulatory frameworks that balance the interests of suppliers and users, in areas such as the protection of intellectual property rights, and digital rights management, without disadvantaging innovative e-business models.[76]

Copyrights in the context of user-created content

Copyright issues related to UCC arise in a number of different ways. At the outset, it may be helpful to distinguish between "original works" created by users and works created by users from pre-existing works (commonly called "derivative works"). Original works identified as UCC raise the same copyright issues as original works created under other circumstances and can present relatively familiar issues of control, commercial exploitation, and protection in the online environment. Derivative UCC works (such as fan fiction or a blog that incorporates some or all of a protected work) highlights a difficult copyright issue, *i.e.* whether such derivative works are acceptable uses permitted by the respective jurisdiction's exceptions and limitations (sometimes referred to as "fair use") or an unlawful infringement of the creator's exclusive rights.

Original works created by users: A large amount of UCC consists in individuals or groups uploading their own original content (*e.g.* photos, videos, art) to their personal blogs or other platforms. The originality requirement to obtain copyright does not necessarily imply an elevated standard of quality or effort (WIPO, 2006a). The creators of works identified as UCC are automatically granted the same exclusive rights as creators in other circumstances are granted. Infringement issues surface when third parties exercise one or more of the UCC creator's exclusive rights without permission or the use is not an exception or limitation (sometimes referred to as "fair use"). In the same vein as for other forms of content creation, copyright for UCC can be considered a catalyst to the production of original works. This holds especially true when creators are interested in pursuing some gainful activity through the commercialisation of their works. Through the control of reproduction and derivative work, these creators also retain control of the way their work is used (including how it is commercialised) and can hence avoid, for example, unwanted modification of their works.

Alongside traditional copyright protection, creators or UCC platforms may – in parallel or in addition – also opt for different licensing schemes, such as the Creative Commons licence. Under these licences others are automatically allowed to copy and distribute a work provided that the licensee credits the author/licensor. In addition, other rights may be reserved or waived (*e.g.* right to create derivatives on non-commercial terms). Examples would be an attribution licence, whereby others can copy, distribute, and remix the work as long as the original author is attributed. While such licensing schemes may permit copying and non-commercial re-use, original authors can specify certain restrictions which have to be observed by those interested in creating derivative works.

Derivative works: Because of copyright law, creators of content identified as UCC, have to respect the exclusive rights of other content producers, *i.e.* of those who choose to work within and those who choose to work outside professional routines and practices (or some combination thereof). Copyright infringement issues may arise whenever someone who is not the copyright holder (or a licensee) exercises an exclusive right, such as adapting the work to create a derivative work, be it for commercial or non-commercial purposes. Copyright issues may thus arise when users create content by using – in part or in full – pieces of others' work without authorisation or where the use does not fall within an exception and limitation. Examples which entail replicating or transforming certain works are the use of particular characters (*e.g.* from Lord of the Rings) in writing fan fiction, using certain images and texts while blogging (*e.g.* using press agency pictures when blogging, using large excerpts of news reporting video footage in one's news commentary), creating lip-synching videos or music mash-ups with samples of existing songs, and the creation of UCC videos while using copyrighted characters, texts or video images.

Copyright laws typically limit in one way or another the copyright holder's ability to control derivative works.[77] Depending on the OECD country, "fair use"- and "fair dealing"- principles and/or specific statutory exceptions allow courts to avoid the rigid application of the copyright statute's exclusive rights when, on occasion, it would discourage creativity, and the public interest in or wide dissemination of knowledge through copyrighted works. Under these circumstances, portions of works can be used without permission and without payment if their use is within one of the copyright exceptions and limitations. Most copyright acts contain limitations for the following activities: personal use, quotation and criticism, comment, parody, news reporting, teaching, scholarship or research, educational and library activities, and – depending on the country in question – other forms of use. In all OECD countries, the latter exceptions are varied reflecting local traditions and are decided on a case-by-case basis. These differences between fair use and copyright limitations are described in Box 7.2.[78] In general, when large portions of a work are taken over or when commercial implications arise, fair use exemptions are less likely to apply (see Gasser and Ernst, 2006).

Box 7.2. Fair use and copyright limitations

Under Article 9(2) of the Berne Convention for the Protection of Literary and Artistic Works and other international copyright treaties[79] signatories are permitted to establish limitations and exceptions at national level but are subject to the so-called "three-step test": The "three-step test" requires that limitations and exceptions must be *i)* confined to special cases, *ii)* not in conflict with a normal exploitation of the work and *iii)* of no unreasonable prejudice to the legitimate interests of the author (see Fiscor, 2002 and Senftleben, 2004 for discussion). The agreed statement concerning the three step test in article 10 of the WIPO Copyright Treaty also underlines that these provisions permit signatory countries to devise new exceptions and limitations that are appropriate in the digital network environment.[80]

National approaches to the determination of exceptions and limitations vary. Rather than specifying a closed list of limitations, common law countries allow for a particular type of limitation on exclusive rights, *i.e.* fair use and/or fair dealing exceptions (Guibault, 1998, WIPO, 2006b). Under US copyright law, for example, a use is permitted if it is determined to be "fair use" as that term is defined by U.S. statutes and case law. US copyright law lists categories of uses that may be fair use under the copyright law, such as criticism, comment, news reporting, teaching, scholarship, and research. This listing is not exhaustive. To determine whether a use is fair, a US court would consider *i)* the purpose and character of the use (*e.g.* use for non-profit educational purposes), *ii)* the nature of the copyrighted work, *iii)* the amount and substantiality of the portion used in relation to the copyrighted work as a whole, and *iv)* the effect of the use upon the potential market of the copyrighted work.[81]

The approach in other OECD countries such as Australia, member countries of the EU, Korea and Japan is rather to define a set of closed purpose-specific exceptions to exclusive rights.[82] In Australia, "fair dealing" is a use of a work specifically recognised as not violating exclusive rights. However, in order to qualify for such exceptions a use must fall within closed purpose-specific exceptions (*e.g.* review or criticism, parody, satire, research or study, news-reporting) and certain circumstances must be met (depends on the nature of the created work, effect of the use on any commercial market for the work, etc.). In the United Kingdom, the "fair dealing" approach also specifies a list of situations where "dealing" with a protected work is permitted.[83] "EU Directive 2001/29/EC on the harmonisation of certain aspects of copyright and related rights in the information society (EUCD)" introduces an exhaustive list of optional exceptions and limitations.[84] This list is amenable to the various legal traditions of the EU member states. The EU Directive also mandates the adherence to the three-step test described above.[85]

Finally, Korean and Japanese Copyright Acts contain categories of uses for which the exclusive rights of authors does not hold: *e.g.* educational uses, quotation, news reporting, etc.[86] In particular, the fair use exception connected to non-profit performance may – under certain circumstances – be relevant to UCC.[87]

No matter whether copyright systems follow a "fair use" doctrine or whether they opt for a specific list of exemptions, broadly speaking the application of these standards is complex and it is difficult to predict what a court will decide when applying them (See Cotter, 2006 and Litman, 2006, for a related discussion). Also, in the digital UCC environment one question is how to adapt the parameters of certain copyright exceptions and limitations, such as fair use, when citations and compilations are increasingly prevalent and easy. In a multi-media environment with mixes of text, video, and graphic works, concepts such as "citation" may be blurry. As with any other use being made of a work still under copyright protection, if no exemption can be invoked, the creator of derivative UCC has to obtain permission from the original authors to create the UCC (*e.g.* for remixes, mash-ups). There remains a degree of legal uncertainty on the side of the creator of the original work as well as with the creator of the derivate work. While this legal uncertainty may lead to the creation of less derivative works, it also has advantages, namely that courts maintain some degree of flexibility when deciding on whether a use is a permissible exception.

The general question for UCC is what are the effects of copyright law on non-professional and new sources of creativity and whether copyright law may need to be examined or does not need to be re-examined, in order to allow coexistence of market and non-market creation and distribution of content and spur further innovation.

Current legal interpretation maintains that standard copyright rules and exceptions are a necessary condition for creativity and that the exceptions and limitations work well (*e.g.* Ginsburg, 2002). In principle, copyright limitations provide ample opportunity for a use to qualify as a permissible exception or limitation. Future case law may determine the boundaries of exceptions and limitations and produce clarity in the UCC context. This is also the thinking pursued in recent national and international legislative approaches (*i.e.* the WIPO Internet Treaties) which propose a combination of exclusive rights and exceptions and limitations. If necessary, existing limitations can also be amended. The Gowers review in the United Kingdom, for example, suggests amending applicable EU copyright law to allow for an exception for creative, transformative or derivative work, within the parameters of the Berne three-step test and to broaden the list of exceptions to copyright for the purpose of caricature, parody and pastiche (UK Treasury, 2006). Overall, it must be clear that a sizeable share of original UCC works is not concerned by these considerations as no derivative works are involved, and many examples of UCC qualify for the standard limitations on copyrights.

On the other hand, proponents of UCC have voiced concerns which are largely based on the idea that non-commercial users have different incentives to create, use, and to share than established professional content holders and that these incentives should be preserved due to their social and cultural impact (see, for example, OECD 2006b, for related discussions and follow-up discussions of the World Summit on the Information Society[88]). These concerns centre on how the copyright law on derivative works could stifle some of the creativity that digital technology enables (Lessig, 2004; Fisher, 2004, 2006). It has been argued that some would-be users are deterred from engaging in conduct that could fall within the ambit of fair use (and hence be legal), due in part to concerns over incurring legal fees and also to the uncertainty and unpredictability of the fair use approach itself (Cotter, 2006). The idea that the IPR system may not have kept pace with progress in this sense and that content production based on the reuse of existing materials – such as sampling or mash-ups – should not be penalised *per se* has been echoed at the policy level.[89]

Facilitating UCC creation: More flexible and efficient licensing processes for copyrights (including for non-UCC areas) have been suggested in the digital context (EU, 2006; OECD, 2006c). Current licensing regimes have been seen by some to be unduly burdensome because of the costs involved or the inability to identify and locate the author of the original work. In some cases the original author of a work will not be identifiable and cannot be contacted, and hence no legal use of the material can be made.[90] Solutions such as new ways to license copyright or new technologies to facilitate licensing could be explored and have – in some cases – been implemented. This could, for example, involve the creation of clearing houses/centres for the attribution of rights to UCC and other creators. From the point of view of commercial copyright interests, any such changes should not be solely to benefit creators of UCC and to the detriment of their commercial interests.

Furthermore, the expansion of fair use-type provisions to derivative works that are more than just copying (*i.e.* that have real transformative and creative value) and that are non-commercial, have also been proposed (Litman, 2006, Fisher, 2004) – often based on the argument that remixing of others' work can also serve to benefit original creators by providing increased exposure. Commercial use of such derivative works would continue to be regulated by the regular statutory rights and limitations.[91]

These suggestions may imply changes to copyright laws, and they rely on a clear dividing line between commercial and non-commercial content, which may be difficult to establish taking into account the diversity of UCC services and related business models. Moreover, the suggested benefits from such new approaches would have to be weighed very carefully against their costs, including, for example, to the established commercial content industry which produces significant economic value. Beyond suggestions, more study is needed of the extent to which UCC creates proven, valuable creative works and associated private and public benefits, as well as of what the potential economic damage is, if any, to the established commercial content industry. So far available statistics seem to demonstrate rapid growth of UCC within current frameworks. One question is to what extent could this growth be even more rapid, whether it comes at the expense of the commercial content industry and other creators, and whether there is a likelihood of constraints on further growth due to difficulties encountered under copyright law.

To date, the attention of rightholders is mostly focussed on UCC platforms which host unmodified snippets or entire reproductions of their original works without authorisation (see section below). So far there seem to be relatively few legal cases directly aimed at the creation of non-commercial derivative works by individuals. However, there are increasing legal actions in the form of take-down notices and "cease and desist" letters which are sent to UCC platforms and individuals asking them to take down certain potentially unauthorised content and which may not reach courts.[92]

Finally, experimentation with new models for the economic use and creation of new digital content is ongoing which does not rely on changes of statutory rights and exceptions of copyright regimes, *e.g.* flexible licensing regimes such as the Creative Commons.[93] The idea is to facilitate the release of creative works under liberal licence terms that would make works available for sharing and reuse. These may address the particularities of content created by amateurs and allow for a parallel coexistence between traditional commercial content and free UCC. But their impacts are not clear and merit further study, including positive or negative effects on the creation process (OECD, 2006a, 2006b). Introducing further diversity in access and licensing regimes to copyrighted works may also have disadvantages (Elkin-Koren, 2006). In sum, the legal meaningfulness of such licences has not yet been fully assessed by research and courts.

Copyrights and the terms of service of UCC sites

As shown by the analysis of the terms of service (Table 7.1), most UCC sites specify that they retain IPRs in their respective content (*e.g.* text, software, graphics, layout, design – especially in cases such as Second Life or social networking sites with their own software and content).

Table 7.1. Intellectual property provisions in terms of service of UCC sites

Content created by site	• Most sites specify that they retain IPRs in their respective content (*e.g.* text, software, graphics, layout, design) under copyright.
Content created by users	• Most sites specify that users who post content retain ultimate ownership, but that they have given the site a licence to use content without payment. In other words, by posting the content the sites receives a limited irrevocable, perpetual, non-exclusive, transferable, fully paid-up, worldwide licence (with the right to sublicense) to use, modify, publicly perform, publicly display, reproduce, and distribute such content.
	• Most sites specify that this licence does not grant the site the right to sell the content, nor to distribute it outside of the respective service.
	• Most sites pledge to mention the identity of the user, the author of the work, and also the title of the work, in so far as technical conditions make this possible.
	• Most sites specify that the licence terminates at the time the user removes his/her content.
	• Some sites reserve the right to prepare derivative works (modify, edit content posted by users) or the right to adapt. At times, it is specified that the site may commercially exploit the works posted by users.
	• Some sites specify that users lose their IPRs and forfeit payment in perpetuity (even when the content is removed). Sometimes the sites also ask the user to admit "moral rights" (meaning that the site does not have to give the author credit).
	• Some sites require the user to agree that content will be subject to the Creative Commons licence.
	• Some sites reserve the right of reproduction, i.e. the right to reproduce, without limitation, on any known or unknown medium, current or future, especially optical, digital, paper, disc, network, diskette, electronic, DVD, etc.
	• Some sites reserve the right to distribute the work or to sublicense rights to third parties. Mostly, it is proposed that revenue from these activities be shared between the user and the site.
	• Some sites reserve the right to use the name and content of users for advertising and promotional purposes (promotional licence).
Reservation to terminate the service	• Most sites reserve the right to modify or terminate the service for any reason, without notice, at any time, which may have consequences on content stored or acquired by users.

Source: OECD based on a review of the terms of service of a sample of 15 widely-used English-speaking UCC sites.

UCC platforms usually grant users who upload content the right to retain copyright in their work. This right is enforceable and applicable both online and offline, both for non-profit and commercial ventures. According to the terms of services of the sample of UCC platforms, users agree that they have given the site a licence to use the content, mostly without payment.[94] Competitors with profit-sharing strategies and arrangements have also emerged. At times the sites reserve the right to prepare derivative works of the content posted by users and the terms of service require the uploader to waive their moral rights. Some sites reserve the right to commercially exploit the works posted by users or to license the content to third parties. Some sites require the user to agree that the content will be subject to the Creative Commons licence. In some cases, unclear terms and conditions or a failure of users to read the latter may lead the user to agree to granting additional rights (even after the user has taken down the content and even for commercial purposes). Often, however, the licence agreed to by the user terminates at the time the user removes such content from the Internet platform site, hence terminating the licence granted to the UCC platform.

A further issue is that some sites reserve the right to modify or terminate the service for any reason without notice at any time, and that this may have consequences on content stored or acquired by users. If, for instance, a UCC site terminates or modifies the service a user may lose his/her uploaded content, the way it was tagged and organised, potentially his/her avatar, and with it many hours of labour and/or money invested to create the content.

Copyrights and the liability of UCC platforms

As discussed above, the growth of UCC is accompanied by the emergence of many sites and online intermediaries hosting the content which users upload. In some ways their existence drives the growth and access to UCC (and vice versa). From a copyright perspective, however, the question emerges in which way online intermediaries are liable for copyright matters.

For example, copyright issues arise when users post unaltered third party content on UCC platforms without authorisation (*e.g.* uploading parts of popular TV series without the explicit consent of the content owner). This activity is outside the scope of UCC as defined in this study, but it is still a key concern of rightholders, who may seek to hold the UCC platforms directly or indirectly liable for copyright infringement. Additionally, posting UCC that is created through the adaptation of pre-existing work may also raise copyright issues for UCC platforms, *e.g.* whether the particular use is permissible under exceptions and limitations such as fair use, and if not permissible, whether the UCC platform is liable for direct or indirect copyright infringement as a result or otherwise exempted from liability for the infringement.

Rightholders are beginning to engage in relevant actions against potential infringement on UCC platforms. Associations representing content owners have sent take-down notices and have asserted potential lawsuits against UCC platforms.[95] An example of interactions between rightholders and a UCC site is the recent legal case between YouTube and the Japanese Society for Rights of Authors, Composers and Publishers (JASRAC) which complains about music videos being uploaded to YouTube without rightholders' permission. Major media companies have also requested online video sites to remove their content.

Some UCC platforms have defended the posting of unaltered third party and alleged infringing derivative UCC content on their platforms by arguing that they are similar to Internet Service Providers (ISPs) who can, under certain circumstances, be exempt from liability for content uploaded by their users (see Litman, 2000). The essential question is whether online intermediaries be treated as electronic publishers, and thus liable for content on their servers (Koelman and Hugenholtz, 2003; WIPO, 2005). As shown in Box 7.3, national legislatures have dealt with the liability of online intermediaries in different ways, which raises issues for internationally operating online intermediaries.

Whether UCC platforms can be treated as a "mere conduit" under exceptions for online intermediaries is an ongoing question. As depicted in Table 7.2, most UCC sites specify that users who post content are responsible for it. They must own all rights to it or have express permissions from the copyright owners to copy and use images. They may not violate or infringe upon the rights of others. Moreover, the terms of service of UCC sites specify that when valid notifications are received, the service provider usually pledges to respond by taking down the unauthorised content.[96] Then the owner of the removed content is contacted so that a counter-notification may be filed.

Table 7.2. Intellectual property provisions and responsibilities in terms of service of UCC sites

Users are responsible for uploaded content	Most sites specify that users who post content are responsible for it. Those uploading must own all rights to it or have express permissions from the copyright owners to copy and use images. They may not violate or infringe upon the copyrights of others.
Take down notice procedure	When valid notifications are received, the service provider usually pledges to respond by taking down the offending content. Under some legal regimes, it is specified that the owner of the removed content is contacted allowing him/her to file a counter-notification.

Source: OECD based on review of the terms of service of a sample of 15 widely-used English-speaking UCC sites.

Box 7.3. Copyright liability of online intermediaries

In their copyright or e-commerce laws many OECD countries have addressed the liability of ISPs and other information intermediaries who merely deliver content by creating liability exceptions ("safe harbour" under the US Digital Millennium Copyright Act[97]) for these entities. This is an exemption from secondary liability but requires the online service providers to remove infringing materials upon notice. In the U.S. Digital Millennium Copyright Act, for instance, following the "notice and take down procedures", ISPs are responsible for taking down unauthorised copyrighted material when a legitimate claim of a rights holder is presented to them.[98] They are also responsible for terminating access by repeat infringers. Similar principles on the liability of online intermediaries also exist in Australian copyright law[99], *i.e.* providers are not obliged to actively self-monitor for infringing activity.

The EU Electronic Commerce Directive 2000/31/EC also establishes an exemption from liability for intermediaries where they play a passive role as a "mere conduit" of information from third parties and limits service providers' liability for other "intermediary" activities such as the storage of information.[100] No general monitoring obligation can be imposed on the service provider.[101] Activities which involve the modification of transmitted information, for instance, do not qualify for this exemption. This EU Directive also encourages hosting services providers to act expeditiously to remove or to disable access to the information concerned upon obtaining actual knowledge or awareness of illegal activities.[102] Such mechanisms are to be developed on the basis of voluntary agreements and codes of conduct between all parties concerned.[103] In addition, in EU Member States such as Italy and France but also on the EU level, public-private partnerships emerged regrouping ISP, rightholders and the government to promote the development of legal online content (OECD, 2005b). Some of the resulting codes of conduct imply that upon notice ISPs should – while respecting privacy laws - contact users uploading infringing material and potentially terminate their accounts.[104]

While under certain circumstances UCC platforms may benefit from the exemption, UCC platforms could also be held liable under domestic law for facilitating, inducing or authorising copyright infringement (recognising that this form of liability is treated slightly differently among OECD member countries). Under the principle of contributory liability, it may be that such online intermediaries are found liable to induce, cause or materially contribute to the infringing conduct of their users. This holds particularly in cases where UCC platforms have knowledge of the infringing activity (*i.e.* "wilful infringement"), when they do not simply host but edit or categorise the content (which is mostly the case), when they induce users to post unauthorised content (*c.f.* the US Supreme Court Ruling vis-à-vis the Grokster case[105]), or when they derive revenues (*e.g.* advertising-related) from unauthorised postings.[106]

In some cases the take-down notice procedure may lead to UCC being taken down without a legitimate reason. UCC platforms receiving notifications from media companies may prefer taking down the respective content rather than risking legal pursuit. Courts are not involved in this decision. There have been cases where UCC has been deleted from UCC platforms by error *e.g.* when the title of the video clip resembles copyrighted content or when in fact, fair use or free speech exceptions may apply.[107] While the individual may have the right to counter-notification in some OECD jurisdictions, it is difficult to obtain information on whether these counter-notifications succeeded in restoring the non-infringing content to the UCC platform.

Despite these concerns rights holders, have also increasingly been interested in deriving value from UCC platforms and in implementing appropriate business models while leaving the copyrighted material on UCC platforms often also noting the significant promotional value of such content. Often legitimate IPR challenges will be resolved through appropriate business agreements between rights holders, UCC platforms and other associated entities. Upon the request of rightholders and to avoid legal actions against them, some UCC platforms have announced or adopted technologies preventing the upload of unauthorised content (*e.g.* acoustic fingerprinting, watermark detection).

Digital rights management

Digital rights management (DRM) has a number of impacts on user-created content and DRM technologies affects UCC in two general ways (the opportunities and challenges raised by DRM and the need for appropriate disclosure are discussed in more detail in e.g. OECD 2005b; 2006c, f). First, DRM can enable digital distribution of UCC just as it has for content that is not identified as UCC. Second, DRM may limit access to works for creators of derivative works or reproductions that are permitted under certain copyright exceptions and limitations, such as "fair use" or other statutory exceptions and limitations.

DRM technologies have been seen as business enablers for the digital distribution of content and drivers for the variety of new business models that consumers may want (OECD, 2005b). DRM may facilitate the creation and dissemination of creative works. Content creators and publishers can use DRM to protect their work from unauthorised downloading, viewing and from the possible creation of derivative works. This potentially encourages the content rightholders to make content available for digitisation and subsequent digital sale. DRM also allows the creation of certain new business models. Through their ability to create diverse access schemes to content, DRMs may enable content to be more tailored to consumer demand (*e.g.* the right to purchase time-limited access to songs) and that may – if prices reflect the level of access – increase consumer choice and satisfaction.

DRM creates opportunities for those users creating content who want to protect their copyrights (*e.g.* avoid copying or the creation of derivative works) and/or commercialise their content. It can be envisaged that users who create very popular content may eventually be interested in entering into commercial agreements with publishers, media companies and various distribution platforms. UCC creators may also like to protect their exclusive rights through DRM but not forego their large dissemination. The content security and amenability to new business models which is made possible by DRM then acts as a facilitator of the growth of UCC.

DRMs or technical access limitations more generally are also reported to have negative effects on the growth of some legitimate UCC as they can generally prevent access to and modification of files. This limits access to works for creators of derivative works or reproductions that are legal under allowable exceptions and limitations (see also UK Treasury, 2006).

The WIPO Internet Treaties require signatory governments to provide "adequate legal protection and effective legal remedies against the circumvention" of technological protection measures like DRMs. These new legal norms make it illegal to circumvent existing technological protection measures to access the content – even if access to that content would in certain cases be covered by exceptions or limitation. For example, if a user wanted to make a parody remix of a film or a teacher make an educational video, the technological protection measures could restrict or prevent the user from extracting the portions of the video to do so, even if using portions of the video were permitted pursuant to copyright exceptions and limitations. Thus in some countries it would effectively be illegal to "circumvent" a technological protection measure (*e.g.* DRM) to access content – even if it falls under fair use or other statutory exemptions described earlier. The question then arises how technical protection measures can be implemented while preserving the balance between exclusive rights and fair use (see also WIPO, 2006b). The Gowers Report in the United Kingdom, for instance, has argued for easier possibilities for users to file complaints relating to DRM and to provide more consumer guidance (including through better labelling) (UK Treasury, 2006).[108]

In some OECD countries, the circumvention of DRMs has recently also been made possible through the introduction of certain exceptions. Recently, for instance, the US Copyright Office created new exemptions to the Prohibition on Circumvention of Copyright Protection Systems for Access Control Technologies (*e.g.* when circumvention is for the purpose of making compilations of portions of those works for educational use in the classroom by media studies or film professors).

As technologies continue to develop that allow more people to create UCC more easily, cheaply, and faster, and as copyright holders continue to explore new business models available through DRM, the potential effects of DRM and the legal protection of circumvention for non-infringing uses may need review in order to maintain the balance between exclusive rights and exceptions and limitations as well as a balance between creator and user interests. Further evidence of cases where fair use has been hampered may be needed as currently there is little analysis in this field (Ginsburg, 2002).

Freedom of expression

The Internet can be seen as an open platform enriching the diversity of opinions (including product reviews), various political and societal debates, the free flow of information and freedom of expression. UCC is in many ways a form of personal expression, and user / creators are engaging in a form of democracy where they can directly publish and enable access to their opinions, knowledge and experience. Preserving this openness and the decentralised nature of the Internet may thus be an important policy objective. Censorship, the filtering of information (including through ISPs or UCC sites themselves), depriving users of the access to certain information or tools for self-expression is in contradiction to this principles.[109] As discussed below, a balance must be struck between freedom of expression and other behaviour *e.g.* the posting of illegal or unauthorised copyrighted content.

Information and content quality

User-created content is produced in a non-professional context outside of traditional media oversight and often without any pecuniary remuneration, and this can have implications for the "quality" of material being posted, admitting that the concept of quality is hard to define and has both subjective and contextual aspects. Content quality problems are of two different but interrelated kinds:

- **Information quality and accuracy**: In the case of blogs, commentary and other UCC forms which refer to facts and figures, the accuracy of content and acknowledgement of sources may not be guaranteed. For example, bloggers do not necessarily have an incentive to check the information that they provide or they may not properly cite sources. The risks associated with inaccurate, defamatory and uncheckable information spreading over the web are seen to be increasingly important. The availability of large amounts of information (some accurate and some not) shifts the responsibility to users to correctly assess information found on UCC sites. Especially, younger users will have to develop the skills to differentiate between incorrect and correct information. This is an entirely new way of approaching information – in the previous media setting information was generally thought to be correct. This is not to say that the information quality on UCC sites is generally poor. In some instances information quality may be high as demonstrated by recent comparisons of Wikipedia and Britannica finding few differences in quality (Giles 2005), and about 50% of Asian users believe content on blogs to be as trustworthy as established media.[110]

Furthermore, while the selection process for content of traditional media outlets may be more organised, it too is not immune from quality problems and the provision of inaccurate information (*e.g.* TV content has also been criticised or wrong information released, even if subsequently usually corrected). Also, a large share of UCC not posted anonymously (*e.g.* personal blogs) can be of very high quality as creators care about their reputation, and have high incentives for accuracy.

- **Content quality**: Many UCC posts would not have passed traditional editorial review or media selection processes (*i.e.* the comparison between Figure 5.2 and Figure 5.1). Content posted on UCC platforms may, for instance, have low technical quality (*e.g.* online videos posted may have poor images). Sometimes the quality of the content itself could objectively or subjectively be judged below standard, but although information accuracy can often be verified by accepted standards, quality of content is hard to define. Content posted by users may be of exceptional value to other users, particularly if it has a personal touch, despite suboptimal technical quality and lack of a story or newsworthiness, and high demand for UCC points to demand for such types of content.

As users are free to choose and often rate content, sources of poor information may not draw many visits. Ways and means to improve the reliability and the quality of information on sites hosting UCC have been developed which may alleviate problems of inaccuracies and onerous oversight. Sites hosting UCC, aggregators and other mechanisms assessing quality and credibility which effectively harness the unbiased "collective intelligence" of users may have greater ability to correct misinformation through "collaborative filtering". Such tools may also find business and other applications to refine large amounts of information.

Different ways of governing UCC sites have emerged: **Pre-production moderation** – content submitted by users is not posted until reviewed by an expert or a person controlling exactness and quality; **Post-production moderation** – content submitted by users is accessible by everybody immediately but moderation may opt to review, make changes or delete the content after posting; **Peer-based moderation** – content submitted by users is available immediately, but can be edited, reviewed or deleted by certain or all users of the same UCC platform. Due to the fact that a larger group of people is involved, peer-based moderation is considered to best maximise the potential of UCC platforms, but this system of moderation also places the greatest responsibility with the user community.[111] New rating and recommendation schemes have also emerged for (*e.g.* social filtering and accreditation, see Kolbitsch and Maurer, 2006).

As the importance of reviews, tags and ratings increases, users may be tempted to abuse those systems by including wrong or biased reviews (review fraud) or engage in misleading tagging of their content (*i.e.* a member uses popular but irrelevant keywords to describe his/her video or other content in order to draw more traffic). This reduces the overall reliability of ratings and searchability of the network. UCC sites make an effort to reduce the possibility of such abuse, but overall, problems of information and content quality and accuracy may remain.

Mature, inappropriate, and illegal content

Few technical limits are imposed on users with respect to their thoughts expressed or to their actions and most UCC platforms allow for relatively free expression (see also the situation of online multiplayer computer games in OECD, 2005d). Sites hosting UCC have been sources of explicit language and behaviour, mature content, gambling, harassment, and defamatory speech. In the United Kingdom and Korea, for example, policymakers have voiced concerns over violent videos on video-sharing sites (*e.g.* students being assaulted and filmed by other students), and video-sharing sites which do not filter content or allow live broadcasts can be a new source of concern. Different OECD countries have different rules, especially as regards indecent or mature content, and the degree to which rules on freedom of expression would permit such expressions or make them unlawful is hard to establish in general. Furthermore, despite concerns regarding the international access to UCC sites, it should be kept in mind that these are one of the many Internet-based sources of such content raising concerns.

Most UCC platforms make it quite clear that they do not police content or that they do not assume editorial responsibility for the content created (see Table 7.3). This is an important point if aiming to enforce laws on illegal content or to reduce the spread of content which may be deemed inappropriate or harmful to certain viewers, *e.g.* minors.[112]

Table 7.3. Content and conduct provisions in terms of service of UCC sites

Content regulation and editorial responsibility	• Most sites specify that users are solely responsible for the content that they publish or display on the website, or transmit to other members. The sites specify that they have no obligation to modify or remove any inappropriate member content, and no responsibility for the conduct of the member submitting any such content. • The sites reserve the right to review and delete or remove any member content which does not correspond to defined standards. • Some sites use age and content ratings or have areas for content which is rated mature.
Community standards	• Most sites have community standards on intolerance (derogatory or demeaning language as to race, ethnicity, gender, religion, or sexual orientation), harassment, assault, the disclosure of information on third parties and other users (*e.g.* posting conversations), indecency, etc.
Actions to enforce standards	• Sites specify penalties when users infringe community standards. They range from warnings, to suspensions, to banishment from the service. The creation of alternative accounts to circumvent these rules is tracked.

Source: OECD based on a review of the terms of service of a sample of 15 widely-used English-speaking UCC sites.

Table 7.4. Age limits and warnings in terms of service of UCC sites

Age limits and age ratings	• Most sites require users either to be 13-14 years old or 18 years. Some put the bar at 16 years. Some have special sub-sites or parts of virtual worlds which are reserved for teenagers.[113] Older users are not permitted to use these sub-sites.
Warnings on releasing information	• UCC sites now post warnings not to post contact details, warnings about adding strangers to friends' lists, warnings about inappropriate content, warnings about posting something which could embarrass the user or somebody else, warnings about reporting false age, warnings about phishing, *i.e.* third parties trying to get personal information, usernames and passwords.

Source: OECD based on a review of the terms of service of a sample of 15 widely used English-speaking UCC sites.

Because of such concerns, many UCC platforms and communities have adopted community standards and associated rules made by the service provider to reduce the incidence of inappropriate content and actions (see Table 7.4). These include, for example, rules on tolerance, on harassment, on assault in virtual worlds (*e.g.* shooting, pushing, etc.), on privacy and the prevention of disclosure of information, on indecency, or on undesired advertising content. If not respected, the service provider reserves the right to take actions against the user (*e.g.* temporary or permanent suspension of accounts). In general, however, it remains difficult for businesses, online communities or governments to monitor all content and to clearly determine what content is illegal. In particular, this is a problem for children's access to the Internet, and although UCC platforms often specify age limits in their terms of service (see Table 7.4) these may be difficult to police.

Technological, legal, self-regulatory solutions may help to limit access to such content. Age rating systems or age limits are seen as important to ensure protection of minors, but these rating systems need to be clear and increasingly internationally recognised and adhered to, in order to be meaningful (see the games study, OECD, 2005d). Filtering software and other parental controls may also provide solutions.

Safety on the Internet and awareness raising

When users create profiles on a particular site there is no verification of the real identity of the user. This can be useful in cases where users may wish to create parodies or political pages, where a certain degree of anonymity may also stimulate creativity. Yet it can also pose a risk when users misrepresent themselves for illegal purposes, and some sites have established greater verification of a user's identify via their school or work e-mail address and limit networks to schools or workplaces.[114]

Thus users may misrepresent their identities, for example by pretending to be a different age or gender and to deceive other, particularly younger, users. There are documented cases of sexual offenders and other criminals gaining access particularly to young or vulnerable users via social networks. It remains to be seen however if these remain relatively isolated cases. As the Internet is an open platform, offenders may also be easier to track online than offline, as evidenced by successful police investigations.

Implementing robust safety measures, educating parents and children, and trying to minimise the risks of such behaviour should be a priority for law enforcement, government officials and social networking sites (Magid and Collier, 2006). Several initiatives have started to educate children and parents (*e.g.* SafeTeens.com, BlogSafety.Com, SaferOnlineDating.org) and technological solutions such as monitoring software are available (Software4parents.com). Despite shortcomings, age limits and rating schemes and software technologies may play an important role. The UCC platforms themselves have started to foster awareness concerning the dangers related to providing private information (see Table 7.4).

Privacy and identity theft

Concerns have been raised about users increasingly posting more information online about their identities, their lives and those of others (*i.e.* friends, family). Users post photos and videos, publicly accessible profiles on social networking sites, and online journals with intimate details of their lives on blogs and sites. While such sites offer privacy settings to limit the availability of this information to personal contacts or friends, many users choose to make their information publicly available. In principle, information which is not displayed publicly is protected and not sold to third parties (Table 7.5). In the case of a merger or acquisition by a third party, however, this information is an asset which is acquired (Table 7.5). There may also be cases of data leakages which could prove particularly damaging, although so far little is known about such cases which may have occurred via UCC sites.

Table 7.5. Privacy provisions in terms of service of UCC sites

Privacy	
	• Most of the sites collect personal information relevant to the service stating that this is to provide the user with a customised and efficient experience. This information is protected and not sold to third parties.
	• Sometimes personal information uploaded on SNS sites is provided to advertisers (sometimes delivered directly) and other parties in a personally identifiable manner and aggregate usage information in a non-personally identifiable manner to present to members more targeted advertising.
	• Most sites reserve the right to transfer personal information in the event of a transfer of ownership or sale of assets.
	• Sites specify that personal information may be released for law enforcement purposes.

Source: OECD based on a review of the terms of service and the privacy policies of a sample of 15 widely used English-speaking UCC sites.

Another potential negative consequence of the vast amount of personal information available online could be increased occurrences of privacy violations or identity theft (phishing). SNS sites are reported to have been used to phish for users' personal information through spam campaigns. Individuals have used UCC platforms to expose content about somebody else (*i.e.* including posting online videos or other content without the consent of the persons involved) or creating accounts on behalf of another person with false information or content.[115] As a result, normal life has been compromised in some documented cases (Sang-Hun, 2006). Other examples exist in which employers have made use of SNS to screen potential employees. Finally, identity thieves can much more easily track down information to mimic someone else's identity. Further work would be useful on privacy challenges resulting from UCC, including users voluntarily making their information public.[116]

Impacts of intensive Internet use

The impacts of intensive Internet use are an emerging concern. While this phenomenon is not particular to UCC, the popularity of these sites has contributed to particularly intensive Internet use. The blurring of the real and the virtual world may lead users to spend large amounts of time on the Internet and they may fail to devote enough time to other obligations (*e.g.* school, work, and even sleeping and eating). Emotional attachment to online friends and activities may lead to a deterioration of relationships outside of the Internet. Research also reveals a growing number of those for whom the medium becomes a consuming habit with potentially negative consequences (Aboujaoude *et al.*, 2006)[117] and whose symptoms are often referred to as "on-line addiction" (Young, 1996; Minoura, 1999; OECD, 2005d).

Only a few OECD countries report data on Internet addiction (see Figure 7.1 for Korea).[118] According to official Korean data related to online gambling in particular, the number of Korean Internet users with possible addiction was 3.3% of Internet users in 2004 and 2% in 2006 (and 11% with suspected addictions).[119] The Korean government has programmes to reduce Internet addiction while educating students on "healthy Internet use" (*e.g.* courses and the designation of Internet-free days).[120]

Overall, it needs to be kept in mind that such intensive use is not particular to user-created content, but relate broadly to how people decide to manage their Internet and other media usage habits (*e.g.* TV, BlackBerry). Users who engage more actively on UCC sites may have previously been watching TV in a passive fashion. Sometimes the relationships created on UCC sites may not be solely virtual as they may subsequently build offline relationships via on-line meetings. With 3D video conferencing and other technological developments emerging, virtual images and Internet-based communication platforms may facilitate day-to-day interactions.

Figure 7.1. Internet "addiction" as reported by the Korean government, 2006

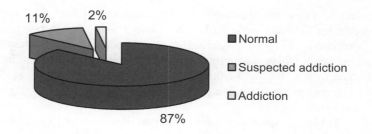

Source: Korean Ministry of Information and Communication.

It is worth stressing that the social impact of the Internet in general and the impact of UCC-related pass-times and communication on society and personal relations have not yet been researched in detail. The spectrum of predictions ranges from Internet communications leading to the "breakdown of personal relationships and social contact" to Internet communications "holding great promises for improving real life relationships and tasks". Recent assessments point to people communicating more than ever but that their pattern of communication and interaction has changed (Statistics Canada, 2006[121]). There is also insufficient understanding of how media consumption generally affects brain processing, learning, attitudes, and behaviour, *e.g.* the impacts of virtual worlds on behaviour, or on learning / skills (see also OECD, 2005d for skills in the context of online games). More research in these fields is warranted.

Network security and spam

Like other information technologies, the participative web is not immune to information security risks. Many Internet sites today serve as platforms for creation and sharing of content. One of the key factors in the growth of UCC was the ease of use in creating and publishing, including CMS, blogging services and wikis. As opposed to web pages of earlier generations, Internet sites now enable the posting of content and the modification of sites. This new interactivity and uploading of content from a large user base can be a source of security problems (Finjan, 2006, Evers, 2006, Nuttall, 2006). In some cases, such sites have been used by hackers who uploaded malicious content (*e.g.* viruses), achieving quicker propagation. In other cases, the greater openness of the platforms for external contributions can cause problems unintentionally. Often, bad implementation of the technology rather than the technology itself may be the root of the problem.

Despite ant-spam provisions in the terms of services of widely used reputable UCC sites (Table 7.6), spam blogs ("splogs"[122]) exist to promote the spammers' site, advertisements or related links, and they include comment spam where a spammer will post comments (often hundreds) to a legitimate user's blog, and wiki spam, where spammers will take advantage of the capacity to rapidly edit a page.[123]

Table 7.6. Spam provisions in terms of service of UCC sites

Spam	• Most sites prohibit illegal and/or unauthorised use of the services, including collecting user names and/or e-mail addresses of members by electronic or other means for the purpose of sending unsolicited email.

Source: OECD based on a review of the terms of service of a sample of 15 widely used English-speaking UCC sites.

Technologies may help to reduce spam.[124] For example, there are tools such as Akismet within blogging software to identify such comment spam. These tools may either automatically delete such comments or put them into a queue, where the blogger can review them. Search engines have sought to combat link-related spam by the implementation of the "no follow" tag attribute.[125] Requiring user registration may also limit spam.

Virtual worlds, property rights and taxation

Increasingly virtual worlds and games are platforms for real economic transactions. Users purchase virtual land and properties, create objects and sell them, and develop the skills and looks of their avatars. Sometimes these commercial transactions take place within the virtual world, sometimes outside of it (*e.g.* selling virtual objects on eBay). Increasingly this phenomenon of virtual economies with real-life impacts is taken seriously (Castranova, 2001, 2005). The question of how to "price" virtual land which in theory is not scarce is just an example of the new questions raised.

On a more practical level, commercial exchanges between the hosting site and the user or between the users themselves are likely to trigger legal disputes as ownership and commercial activity increases. As objects created by the user are often inextricably tied to the virtual world itself, establishing straightforward ownership rights may be difficult. Disputes have already arisen when the hosting site terminated a user's account, thereby deleting objects and land, for actions on the UCC platform banned by the terms of service (see last row of Table 7.1).

Commercial activity in or around virtual world content is also increasing interest by tax authorities.[126] Growing virtual assets and capital gains can be translated into off-line economic gains (*e.g.* when users convert money back to real money or when they sell virtual items on consumer-to-consumer exchanges such as eBay), and tax authorities are beginning to investigate how such transactions should be treated for taxation.[127] Tax authorities will be dependent on the taxpayer actually declaring such sales as income to avoid an electronic version of the "underground" economy. Finally, a number of OECD countries levy wealth and inheritance taxes. Interesting questions arise as to how these taxes apply to "virtual", intangible wealth and thus unrealised gains. At present, tax authorities' attention has not been drawn to these issues due to a combination of high thresholds for wealth taxes and the relatively low value of virtual wealth, but this could become an issue over time.

In the light of the growing influence of virtual worlds, governments will increasingly be faced with associated regulatory questions, be it in the tax or other legal areas, especially when sites operate in a legal vacuum or when it is unclear if and how existing laws apply in such online environments.

Governments as producers and users of content

Governments posting content to inform their citizens are part of e-government policies and does not relate directly to user-created content. Nonetheless, governments may encourage comments or discussions from citizens via discussion platforms at the local (virtual town hall), regional or national level or via other Internet-based tools. The latter may either inform ongoing debates relating to public projects (public constructions, schools, etc.) or they may constitute an outlet for citizens to express themselves, potentially creating greater social cohesion and identification. This area is developing rapidly as public authorities are increasingly turning to new ways of interacting with citizens to increase their own efficiency and be more pro-active in their citizen relations, and is the subject for ongoing analysis.

Conceptualisation, classification and measurement

In general comparative international data on digital content products and industries is not available.[128] Benchmarking the impacts of digital content policies is complicated by the absence of this data. The lack of reliable official data on UCC and more knowledge on changing usage are a challenge. As a result, it is often hard to accurately assess the statistical, economic, and societal effects of UCC. In particular, the social impacts of increased Internet use deserve greater attention. However, OECD has recently proposed ways of improving official data collection on UCC in particular (see OECD, 2007b)

New Internet usage measurement techniques developed by private data services and based on very large sample sizes to monitor online behaviour (sometimes for advertisement-related purposes) provide more detailed data concerning Internet user behaviour, often for targeted advertising, but this also raises challenges regarding the use of that data in the context of the protection of privacy.

Annex

Annex Box 1. Participative web technologies

- *Tagging* is the association of particular keywords with related content, such as photo tagging on photo sharing sites or link tagging on collaborative news sites. Generally, a user will choose a brief selection of keywords that best indicate the content of a particular piece of audio, video, or text. Tagging has played a significant role in social bookmarking sites where users collaboratively store and publish their favourite links.

- *Group rating* and aggregation occurs on sites where users submit links and descriptions of articles and other content and where other users can rate the content. Recommendation engines, particularly popular for music, are technologies enabling users to share tastes and discover new content. An example would be recommendation engines based on musical similarities, or on patterns emerging between users (*e.g.* those who liked x also liked y), or a combination of both.

- *Syndication and aggregation* of data in RSS/Ajax/Atom and other content management systems (CMS). These include:

  - RSS: Really simple syndication is a technology that enables distribution and subscription to content so that users may automatically receive new posts and updates. RSS plays an extremely important role in blogging, and it is increasingly used for videocasting, podcasting, and photo streams. RSS files, also called feeds, transmit structured data which typically include headlines, dates, authors, content summaries and links to the full versions (Bowman, 2003; Gill, 2005). Users can subscribe to a feed and transform the transmitted data into information via a RSS reader. Content creators are able to easily syndicate content for RSS readers. Often, RSS tools are already integrated in publishing software. Readers are able to personalise web services: they do not have to check web pages regularly for new entries but are kept informed by their RSS readers.

  - Atom: The Atom Syndication Format is an XML language used for web feeds. Web feeds allow software programs to check for updates published on a web site. To provide a web feed, a site owner may use specialised software (such as a content management system) that publishes a list (or "feed") of recent articles or content in a standardised, machine-readable format. The feed can then be downloaded by web sites that syndicate content from the feed, or by feed reader programs that allow Internet users to subscribe to feeds and view their content. The development of Atom was motivated by the existence of many incompatible versions of the RSS syndication format.

.../...

Annex Box 1. Participative web technologies *(continued)*

– Ajax (Asynchronous JavaScript and XML) – a Rich Internet Application technique - is a web development technique for interactive web applications which encompasses different technologies. It can be better described as a pattern than a technology — it identifies and describes a particular design technique (McCarthy, 2005). The main advantage of this technique is that "web pages are dynamically updated without a full page refresh interrupting the information flow" and allows creating "richer, more dynamic web application user interfaces". This can be achieved by an Ajax engine which is interposed between the user and the server.

• *Application Mash-ups and Open Application Programming Interfaces (APIs):* Along with audio and video mash-ups, the term can also refer to a combination of multiple web applications. Mash-ups are interactive web applications that draw upon content retrieved from external data sources to create entirely new services (Merrill, 2006). This type of web-based integration aggregates and combines third-party data. API is the interface that a computer system, library or application provides to allow requests for services by other computer programs, and/or to allow data to be exchanged between them (Wikipedia, 2006e). An Open API is available to use free of charge. A variety of web applications uses open APIs, such as Google maps. This has enabled programmers to create combinations, or mash-ups, of Google Maps with other information sources. Examples include a map where all housing ads from Craig's List are placed on a Google Map with relevant information, or the plotting of all of a city's crime incidences on a map with the time and date of occurrence. Other web application mash-ups include video and photo mash-ups, where designers mash photos or video with other information that can be associated with the attached metadata (*i.e.* tags) (Merrill, 2006). An example includes taking news headlines and displaying photos tagged with the particular words. News and RSS feed mash-ups such as NetVibes and My Yahoo aggregate various feeds and present them on a website, enabling users to create a personalised page.

• *Filesharing networks:* Peer-to-peer (P2P) networks are communication structures in which individuals interact directly through decentralised information exchange without going through a centralised system or hierarchy and centralised information control. With P2P technology, users may share information, contribute to shared projects or transfer files (OECD, 2004a). P2P networks provide opportunities for commercial and non-commercial content production and delivery, and providers of content, Internet services and technology are increasingly looking to legitimately "monetise" P2P networks rather than leaving them solely for unauthorised downloading of copyrighted works (OECD, 2006a, 2006b; EITO, 2006).

Annex Box 2. User-created content in China: Video

UCC has started to play an important role in China. Online video provides users with an outlet to express their creativity. A popular video style involves spoofs, or parody-style remixes of other videos. An example was a video which spliced together other videos to make it look as if China had won the 2006 football World Cup.

Top video sharing sites in China (August 2006):

1. Toodou.com

2. Qyule.com

3. Pomoho.com

4. 56.com

5. 365cast.com

Toodou, the country's most popular video sharing, multimedia podcasting and social network site estimated it had 7-10 million unique visitors per month and approximately 50 000 people creating videos mid-2006 (reported 4 million unique users a day mid-2007, and a 47% share of time Chinese Internet users spend on all video sharing websites). Video sharing sites from OECD countries are also trying to take foothold in China. Established Chinese media companies are looking to forge partnerships with online video companies, to position themselves in the next generation of media.

Self-censorship generally occurs on Chinese online video sites, with explicit content and content critical of the government prevented from being posted. Many online video providers employ monitors that view videos and determine the suitability according to self-imposed guidelines for the sites. A significant number of users post their videos anonymously, with fame not necessarily the number one priority. Chinese video websites and clips may soon require approval for posting and distribution from China's State Administration of Radio, Film, and Television (SARFT).

Source: OECD interviews, Pacific Epoch statistics at www.pacificepoch.com and http://digitalwatch.ogilvy.com.cn/en/?p=56.

Annex Table 1. Growth in contributors to Wikipedia

	Jan. 2001	Jan. 2002	Jan. 2003	Jan. 2004	Jan. 2005	Jan. 2006	June 2006
Contributors (min. of 10 total contributions)	10	512	2 423	10 883	50 281	145 564	240 062
Active contributors (min. of 5 contributions per month)	9	205	834	3 202	13 301	47 624	65 905
Very active contributors (min. of 100 contributions per month)	n/a	29	187	684	2 292	7 516	9 142

Source: OECD based on Wikipedia 4 August 2007 at http://stats.wikimedia.org/EN/, and presentation of Frieda Brioschi (Wikipedia) at www.oecd.org/dataoecd/15/14/36133622.pdf.

Annex Table 2. Podcast categories rating

Category	No. of podcasts
News/ politics	11 409
Music	10 342
Religion & spirituality	9 886
Art	7 710
Society & culture	7 207
Education	6 039
Technology	5 878
TV & film	5 671
Comedy	5 106
Business	3 769
Sports & recreation	3 266
Health	2 074
Games & hobbies	1 812
Kids & family	1 301
Science & medicine	1 085
Government & organisations	423
Total	**82 978**

Source: Apple iTunes and "iTunes Podcast Count over 82,000", in: Digital Podcast (9 October 2006), http://typicalmacuser.com/wordpress/?p=134.

Bibliography

ABI Research (2006), "Broadband Video ASPs White Label Platform Providers for Internet TV", Research Brief RB-DMDM-101, available at: www.abiresearch.com.

Aboujaoude, E., L.M. Koran and N. Gamel (2006), "Potential markers for problematic Internet use: A telephone survey of 2,513 adults", in *CNS Spectrums,* No. 11 (10), pp. 750-755.

Anderson, C. (2004), "The Long Tail", *Wired Magazine*, October, available at: www.wired.com/wired/archive/12.10/tail.html.

Balkin, J. (2004), "Digital Speech and Democratic Culture: A Theory of Freedom of Expression for the Information Society", *New York University Law Review*, 79, 1.

Benkler, Y. (2006), *The Wealth of Networks*, available at: www.benkler.org/Benkler_Wealth_Of_Networks.pdf.

Castranova, E. (2001), "Virtual Words: A First-Hand Account of Market and Society on the Cyberian Frontier", *CESifo Working Paper* No. 618, December, available at: http://ssrn.com/abstract=294828.

Castranova, E. (2005), *Synthetic worlds*, University Of Chicago Press, Chicago.

China Internet Network Information Center (CNNIC) (2006), "Statistical Survey Report on the Internet Development in China", The 18th Survey Report, July 2006, available at: www.cnnic.net.cn/download/2006/18threport-en.pdf.

Consumer Electronics Association (CEA) (2006), *U.S. Consumer Electronics Sales and Forecasts Report.*

Cotter, T. F. (2006), "Fair Use and Copyright Overenforcement", Minnesota Legal Studies Research Paper No. 06-69, available at SSRN: http://ssrn.com/abstract=951839.

Crampton, T. (2006), "France's mysterious embrace of blogs", *International Herald Tribune*, 27 July, available at: www.iht.com/articles/2006/07/27/business/blogs.php.

Elkin-Koren, N. (2006), "Creative Commons: A Skeptical View of a Worthy Pursuit", in *The Future of the Public Domain*, Guibault, L. and P. B. Hugenholtz,(Eds), Kluwer Law International, available at SSRN: http://ssrn.com/abstract=885466.

e-Marketer (2006a), "US Online Ad Spending: Peak or Plateau?", available at: www.emarketer.com/Reports/All/Em_ad_spend_oct06.aspx.

e-marketer (2006b), "User-Generated Revenue?" available at: www.emarketer.com/eStatDatabase/ArticlePreview.aspx?1004192.

European Commission (2006), "Interactive content and convergence: Implications for the information society", Study for the European Commission, DG Information Society and Media, October, available at: http://ec.europa.eu/information_society/eeurope/i2010/docs/studies/interactive_cont ent_ec2006_final_report.pdf.

European Union, i2010 High Level Group (2006), "The Challenges of Convergence", Discussion paper, December, available at: http://ec.europa.eu/information_society/eeurope/i2010/docs/i2010_high_level_grou p/i2010_hlg_convergence_paper_final.pdf.

European Information Technology Observatory (EITO) (2006), "Peer-to-Peer (P2P) networks and markets", Part 2, European Information Technology Observatory 2006, Berlin.

Evers, J. (2006), "Blog feeds may carry security risk", CNET, 4 August, available at: http://news.com.com/2100-1002_3-6102171.html.

Finjan (2006), "Web Security Trends Report", 11 October, San Jose, California, USA.

Fiscor, M. (2002), "How Much of What? The Three-Step Test and Its Application in Two Recent WTO Dispute Settlement Cases", *Revue internationale du droit d'auteur,* 192, 231-42.

Fisher, T. (2004), *Promises to keep. Technology, Law, and the Future of Entertainment*, Stanford University Press.

Fisher, T. (2006), speech at the OECD-Italian Minister for Innovation and Technologies Conference, *The Future Digital Economy: Digital Content Creation, Distribution and Access*, Rome, Italy, 30-31 January, available at: www.oecd.org/dataoecd/16/44/36138608.pdf.

Gasser, U. and S. Ernst (2006), "From Shakespeare to DJ Danger Mouse: A Quick Look at Copyright and User Creativity in the Digital Age", Berkman Center for Internet and Society Research, Publication No. 2006-05, June.

Giles, J. (2005), "Internet encyclopaedias go head to head", *Nature*, 15 December, available at: www.nature.com/nature/journal/v438/n7070/full/438900a.html.

Gill, K. (2004), "How Can We Measure the Influence of the Blogosphere", available at: http://faculty.washington.edu/kegill/pub/www2004_blogosphere_gill.pdf.

Gill, K. (2005), "Blogging, RSS and the Information Landscape: A Look at Online News", available at: http://faculty.washington.edu/kegill/pub/gill_www2005_rss.pdf.

Ginsburg, J. C. (2002), "Essay - How Copyright Got a Bad Name For Itself", *Columbia Journal of Law and the Arts*, Vol. 26, No. 1, October 18, available at: http://ssrn.com/abstract=342182.

Gordon, W. J. (1982), "Fair Use as Market Failure: A Structural and Economic Analysis of the "Betamax" Case and its Predecessors", *Columbia Law Review*, 82, no. 8, pp. 1600-1657.

Guibault, L. (1998), Discussion paper on the question of Exceptions to and limitations on copyright and neighbouring rights in the digital era, Council of Europe, Strasbourg, October, MM-S-PR (98) 7, available at: www.ivir.nl/publications/guibault/final-report.html.

Hugenholtz, P. B. (1997), "Fierce Creatures. Copyright Exemptions: Towards Extinction?" keynote speech, IFLA/Imprimatur Conference, "Rights, Limitations and Exceptions: Striking a Proper Balance", Amsterdam, 30-31 October, available at: www.ivir.nl/publications/hugenholtz/PBH-FierceCreatures.doc.

International Business Machines (2007), "Navigating the media divide, Innovating and enabling new business models", IBM Institute for Business Value (February).

Jung-a, Song (2006), "Korean site tackles might of MySpace", *Financial Times*, 31 August, available at: www.ft.com/cms/s/eacfbf3c-3938-11db-a21d-0000779e2340.html.

Koelman, K. and B. Hugenholtz (1999), "Online service provider liability for copyright infringement", WIPO workshop on service provider liability, Geneva, December 9 and 10, World Intellectual Property Organisation Geneva, available at: www.ivir.nl/publicaties/hugenholtz/wipo99.pdf.

Kolbitsch, J. and H. Maurer (2006), "The Transformation of the Web: How Emerging Communities Shape the Information we Consume", *Journal for Universal Computer Science*, available at: www.jucs.org/jucs_12_2/the_transformation_of_the/jucs_12_02_0187_0214_kolbitsch.html.

Lenhart, A. and M. Madden (2005), "Teen Content Creators and Consumers", Pew Internet & American Life Project, 2 November, available at: www.pewInternet.org/pdfs/PIP_Teens_Content_Creation.pdf.

Lenhart, A. and S. Fox (2006), "Bloggers: A portrait of the Internet's new storytellers", Pew Internet & American Life Project, 19 July, available at: www.pewInternet.org/pdfs/PIP%20Bloggers%20Report%20July%2019%202006.pdf.

Lenhart, A. and M. Madden (2007), "Social Networking Websites and Teens: An Overview", Pew Internet & American Life Project, 3 January, available at: www.pewinternet.org/pdfs/PIP_SNS_Data_Memo_Jan_2007.pdf.

Lessig, L. (2004), *Free culture*, The Penguin Press, New York.

Litman, J. (2000), *Digital copyright: protecting intellectual property on the Internet*, Prometheus Books, Amherst.

Litman, J. (2006), "Lawful personal use", paper #06-004, John M. Olin Center for Law & Economics, University of Michigan, available at: http://ssrn.com/abstract=926575.

Magid, L. and A. Collier (2006), "MySpace Unraveled", available at: www.myspaceunraveled.com/

Matlack, C. (2005a), "Let Them Eat Cake – And Blog About It", *Business Week*, 11 July, available at: www.businessweek.com/magazine/content/05_28/b3942082_mz054.htm.

Matlack, C. (2005b), "The Podcast Shaking Up French Politics", *Business Week*, 27 December, available at: www.businessweek.com/bwdaily/dnflash/dec2005/nf20051227_3765_db039.htm.

Mayer-Schoenberger, V. and J. Crowley (2005), "Napster's Second Life - The Regulatory Challenges of Virtual Worlds", Harvard Kennedy School Faculty Working Paper, 28 July, available at: http://ksgnotes1.harvard.edu/Research/wpaper.nsf/rwp/RWP05-052.

Merrill, D. (2006), "Mashups: The new breed of Web app", IBM developerWorks, updated 16 October, available at: www.ibm.com/developerworks/library/x-mashups.html.

Ministry of Information and Communication (2005), *Survey on the Computer and Internet Usage*: Executive Summary, National Internet Development Agency of Korea, available at: http://isis.nic.or.kr/report_DD_View/upload/user_sum_eng_200512.pdf

Ministry of Internal Affairs and Communication (MIC) (2006), White paper on *Information and Communications in Japan 2006*, Chapter 1: Ubiquitous Economy, available at: www.soumu.go.jp/joho_tsusin/eng/whitepaper.html.

Minoura, Y. (1999), Human Development and Education in Digital Revolution, Paper 6, September, Ochanomizu University, available at: www.childresearch.net.

National Computerization Agency of Korea (NCA) (2006), *2006 Informatization White paper*, NCA, Seoul, Korea.

National Internet Development Agency of Korea (2006), Korea Internet White Paper 2006, available at www.mic.go.kr/eng/index.jsp.

Nutall, C. (2006), "The hidden flaw in Web 2.0", *Financial Times*, 8 August, available at: www.ftd.de/karriere_management/business_english/104483.html.

OECD (2004a), *OECD Information Technology Outlook 2004,* Chapter 5, "Digital Delivery", OECD, Paris, available at: www.oecd.org/sti/ito.

OECD (2004b), "The Implications of Convergence for Regulation of Electronic Communications", Directorate for Science, Technology and Industry,

DSTI/ICCP/TISP(2003)5/FINAL, available at:
www.oecd.org/dataoecd/56/24/32983964.pdf.

OECD (2005a), *Science, Technology Industry Scoreboard*, OECD, Paris.

OECD (2005b), "Digital Broadband Content: Music", Directorate for Science,
Technology and Industry, DSTI/ICCP/IE(2004)12/FINAL, available at:
www.oecd.org/sti/digitalcontent.

OECD (2005c), "Digital Broadband Content: Scientific Publishing", Directorate for
Science, Technology and Industry, DSTI/ICCP/IE(2004)11/FINAL, available at:
www.oecd.org/sti/digitalcontent.

OECD (2005d), "Digital Broadband Content: The Online Computer and Video Game
Industry", Directorate for Science, Technology and Industry,
DSTI/ICCP/IE(2004)13/FINAL, available at: www.oecd.org/sti/digitalcontent.

OECD (2005e), "Digital Broadband Content: Mobile Content – New Content for New
Platforms", Directorate for Science, Technology and Industry,
DSTI/ICCP/IE(2004)14/FINAL, available at: www.oecd.org/sti/digitalcontent.

OECD (2006a), *OECD Information Technology Outlook 2006*, OECD, Paris, available
at: www.oecd.org/sti/ito.

OECD (2006b), OECD-Italian Minister for Innovation and Technologies Conference,
The Future Digital Economy: Digital Content Creation, Distribution and Access,
Rome, Italy, 30-31 January, available at:
www.oecd.org/sti/digitalcontent/conference.

OECD (2006c), "Digital Broadband Content: Digital Content Strategies and Policies",
Directorate for Science, Technology and Industry, DSTI/ICCP/IE(2005)3/FINAL,
available at: www.oecd.org/sti/digitalcontent.

OECD (2006d) "Multiple play. Pricing and policy trends", Directorate for Science,
Technology and Industry, DSTI/ICCP/TISP(2005)12/FINAL, available at:
www.oecd.org/dataoecd/47/32/36546318.pdf.

OECD (2006e), "Policy considerations for audio-visual content in a multiplatform
environment", Directorate for Science, Technology and Industry,
DSTI/ICCP/TISP(2006)3/FINAL, available at:
www.oecd.org/dataoecd/21/41/37868139.pdf.

OECD (2006f), "Report on Disclosure Issues Related to the Use of Copy Control and
Digital Rights Management Technologies", DSTI/CP(2005)15/FINAL, available at:
www.oecd.org/dataoecd/47/31/36546422.pdf.

OECD (2007a), *OECD Communications Outlook 2007,* OECD, Paris.

OECD (2007b), "Measuring user-created content: Implication for the "ICT access and
use by households and individuals surveys", DSTI/ICCP/IIS(2207)3/FINAL.

Ofcom (2006), *The Communication Market 2006*, Office of Communications, London.

O'Reilly, T. (2005), "What Is Web 2.0. Design Patterns and Business Models for the Next Generation of Software", 30 September, available at: www.oreillynet.com/lpt/a/6228.

Rainie, L. and M. Madden (2005), "Podcasting catches on", Pew Internet & American Life Project, 3 April, available at: www.pewinternet.org/pdfs/PIP_podcasting2005.pdf

Regner, T., J. A., Barria, J. Pitt, and B. Neville (2006), "Is Copyright Suitable for User-Generated Content? An Alternative Approach", October 2006, http://ssrn.com/abstract=936873.

Sandoval, G. (2006a), "Microsoft lands Facebook ad deal", CNet News.com, 22 August, available at: http://news.com.com/2100-1024_3-6108514.html.

Sandoval, G. (2006b), "CNN Snatching Page out of YouTube's Book", CNET News.com, 30 July, available at: http://news.com.com/CNN+snatching+page+out+of+YouTubes+book/2100-1025_3-6100139.html.

Sang-Hun, Choe (2006), "In South Korea, online rumours can hit hard", *International Herald Tribune*, 14 August, available at: www.iht.com/articles/2006/08/14/news/korea.php.

Senftleben, M. (2004), "Copyright, Limitations and the Three-Step Test - An Analysis of the Three-Step Test in International and EC Copyright Law", Kluwer Law International, The Hague/London/New York.

Shimo, A. (2006), "Living it up in a parallel world", *Globe and Mail*, 29 July, available at: www.theglobeandmail.com/servlet/story/LAC.20060729.SECOND29/TPStory/Entertainment.

Sifry, D. (2006a), "State of the Blogosphere April 2006", 17 April, available at: www.sifry.com/alerts/archives/000432.html.

Sifry, D. (2006b), "State of the Blogosphere August 2006", 7 August, available at: www.sifry.com/alerts/archives/000436.html.

Statistics Canada. (2006), "Our Lives in Digital Times", prepared by Sciadas, G., Science, Innovation and Electronic Information Division (SIEID), available at: www.statcan.ca/english/research/56F0004MIE/56F0004MIE2006014.pdf.

UK Treasury (2006), "Gowers Review of Intellectual Property", Chancellor of the Exchequer, United Kingdom, December.

Van Duyn, A. (2006), "Web ads sector lacks experienced staff", *Financial Times*, 29 August, available at: www.ft.com/cms/s/e2a439cc-378b-11db-bc01-0000779e2340.html.

Van Duyn A. and R. Waters (2006), "Google in $900m ad deal with MySpace", *Financial Times*, 7 August, available at: www.ft.com/cms/s/17e8e67e-2660-11db-afa1-0000779e2340.html.

Von Hippel, E. (2005), *Democratizing Innovation*, MIT Press, London.

VTT Technical Research Centre of Finland (2007), *Googlen mainokset ja muita sosiaalisen median liiketoimintamalleja* ["Ads by Google" and other social media business models], Espoo 2007, Kangas, P., S. Toivonen and A. Bäck (Eds), Research Notes 2369, available at: www.vtt.fi/publications/index.jsp.

Wikipedia (2006), Wikepedia Statistics, available at: http://stats.wikimedia.org/EN/Sitemap.htm, accessed 31 July.

WIPO (2003), "WIPO Study on limitations and exceptions of copyright and related rights in the digital environment", prepared by Ricketson, S., Standing Committee on Copyright and Related Rights, Ninth Session, Geneva, 23-27 June.

WIPO (2004), "WIPO Study on Current Developments in the Field of Digital Rights Management", SCCR/10/2.Rev., prepared by Cunard, J. P., K. Hill and C. Barlas.

WIPO (2005), "Online Intermediaries and Liability for Copyright Infringement", prepared by Waelde, C. and L. Edwards, WIPO Seminar on Copyright and Internet Intermediaries, Geneva, 18 April.

WIPO (2006a), "Trademarks and their relation with literary and artistic works", SCT/16/5, Standing Committee on the Law of Trademarks, Sixteenth Session, Geneva, 13-17 November.

WIPO (2006b), "WIPO Study on Automated Rights Management and Copyright Exceptions and Limitations", SCCR/14/5, prepared by Garnett, N.

Young, K. S. (1996), "Internet Addiction: The Emergence of a New Clinical Disorder", *Cyber Psychology and Behaviour*, Vol. 1 No. 3, pp. 237-244.

Notes

[1] See *OECD Information Technology Outlook 2006,* Chapter 7 (OECD, 2006a).

[2] For an overview of the digital content work programme of the OECD Working Party on the Information Economy see www.oecd.org/sti/digitalcontent; for the January 2006 Rome digital content conference see: www.oecd.org/sti/digitalcontent/conference.

[3] The participative web is a much broader phenomenon than user-created content. Participative web technologies allowing for interaction, blogging and other activities have been particularly conducive to the creation of UCC. But definitions of the participative web (see *e.g.* O'Reilly, 2002, 2005) typically include broader developments, including the rise of new commercial web services or other commercial ventures.

[4] UCC is referred to as consumer-generated media in publications from Japanese official sources, see www.johotsusintokei.soumu.go.jp/whitepaper/eng/WP2006/chapter-1.pdf.

[5] Wikipedia at http://en.wikipedia.org/wiki/User_generated_content, accessed 27 August 2007.

[6] Posting chat messages or using file-sharing services *per se* do not necessarily qualify as UCC. The distinction between UCC (*e.g.* a blog) and just any type of chat content is difficult and subjective. File-sharing sites can be used for the exchange of UCC as well as other types of (copyrighted) content, sometimes without the authorisation of copyright holders.

[7] For a survey of French blogging demographics see Credoc (2006), '*La diffusion des technologies de l'information dans la société française*', available at: www.arcep.fr/uploads/tx_gspublication/etude-credoc2006.pdf.

[8] According to the survey, for girls social networking sites are primarily places to reinforce pre-existing friendships; for boys, the networks also provide opportunities for flirting and making new friends. See 'Microsoft survey: Blogging Phenomenon Sweeps Asia, According to New Research from Windows Live Spaces', in: *Xinhua-PRNewswire,* 28 November 2006, based on Microsoft surveys of its MSN and Windows Live Online Services Business.

[9] Available at www.oecd.org/dataoecd/15/17/36133687.pdf.

[10] See: www.soumu.go.jp/joho_tsusin/eng/Releases/Telecommunications/pdf/news050517_2_1.pdf.

[11] Data on web traffic are an up-to-date and increasingly reliable source of usage information (see OECD, 2004a, Chapter 5). See also the presentation of David Day, Managing Director, EMEA Nielsen//NetRatings at the OECD-Italian Minister for Innovation and Technologies Digital Content Conference available at: www.oecd.org/dataoecd/16/16/36134913.pdf.

[12] Alexa traffic rankings at www.alexa.com/site/ds/top_500.

[13] Remarks of Tony Perkins at 'The AlwaysOn Network's On Hollywood 2006' conference in May 2006. See 'Future of Entertainment: Democratic party', in: *Hollywoodreporter* (26 September 2006) available at: www.hollywoodreporter.com/hr/content_display/tools_data/media_analyst_corner/e3i%2 FWNldeNUbIvVQ4t3MKmReg%3D%3D.

[14] Japan leads the OECD in fibre-to-the-premises with 6.3 million fibre subscribers in June 2006, *OECD Key ICT indicators*. See www.oecd.org/sti/ictindicators.

[15] See presentation of Creative Commons to the OECD-Italian Minister for Innovation and Technologies Digital Content Conference available at: www.oecd.org/dataoecd/15/31/36134387.pdf.

[16] Data from Interpublic's Emerging Media Lab. According to Nielsen//NetRatings, men are 20% more likely to visit YouTube than women. See http://netratings.com/pr/pr_060721_2.pdf.

[17] 'U.S. venture investors betting on energy, Web 2.0', in *Reuters,* 23 October 2006; 'Venture capital investors wake up to Web 2.0' in *Financial Times*, 23 July 2007, available at: http://www.ft.com/cms/s/e2f648d6-38b4-11dc-bca9-0000779fd2ac.html.

[18] Every Creative Commons licence allows others to copy and distribute a work provided that the licensee credits the author/licensor. In addition, the Creator/Licensor may apply different conditions (Non commercial, No Derivatives, Share Alike – the latter allowing you to alter, transform or build upon the work while sharing the resulting work under the same licensing). Licensed content can be explored, search is improved and re-use is promoted.

[19] Now known as Kodak EasyShare Gallery.

[20] Remixing can take the form of "mash-up", whereby two or more songs are edited together. Another related derivative use is sampling, whereby snippets of a song or other audio file are taken and added to another, often in a modified form. An example would be taking Martin Luther King's "I have a dream" speech, rearranging it, and setting it alongside music.

[21] See www.davidbowie.com/neverFollow/.

[22] The Phantom Edit is an alternate version of writer-director George Lucas' "Star Wars I: The Phantom Menace". In addition, user-created videos may also involve animation, where users create or remix animated material. Further, "machinima" is UCC where characters and stories are created within computer games, recorded, and then posted online as short films.

23 In many OECD countries online shoppers consider ratings and reviews a key element in their research when shopping for a new vehicle.

24 Honda has a sponsored site on the blogging network 2Talk About, where users can give their views on the company's products.

25 See, for example, www.bullpoo.com/explore/.

26 See the debate 'User-generated Content - What Does it Mean for Consumers and Marketers' as part of the FTC's hearings on Protecting Consumers in the Next Tech-ade! November 2006 available at: http://ftcchat.us/blog/?p=56#more-56.

27 The difficulties measuring blogs are discussed in OECD (2006b), Chapter 7.

28 See for more details OECD (2006b), Chapter 7. The number of Japanese and Korean blogs is disproportionally large compared with Japanese and Korean general use of the Internet.

29 "Microsoft Windows Live survey: Blogging Phenomenon Sweeps Asia", in: *Xinhua-PRNewswire,* 28 November 2006, based on Microsoft's MSN and Windows Live Online Services Business.

30 Writely and Writeboard enable users to collaborate on documents in a word processing-based environment.

31 Digg had nearly 450 000 users mid-2006, with 30% of frontpage stories coming from the Top 10 users. See http://blogs.zdnet.com/web2explorer/?p=250.

32 Similarly, OpenBC/Xing, LinkedIn and Spoke are professional networking services that attempt to create a networking site or platform for experts and business partners.

33 See http://video.google.com/videoplay?docid=-5182759758975402950 for a more detailed and technical explanation of Second Life.

34 See www.SecondLife.com and Shimo (2006). It is reasonable to assume that these data do not refer to unique users. Individual users can create multiple accounts for different residents and register virtual avatars but never or rarely use them, and thus these data may overestimate the unique audience.

35 Google, for instance, also owns the social networking site Orkut and the wiki-builder Jotspot.

36 See *e.g.* http://wikimediafoundation.org/wiki/Personal_Appeal.

37 In the case of Flickr, 'pro' means an ad-free service for providers of photos, as well as unlimited bandwidth, storage and other features. Feedburner, a widely used blogging software provider, offers additional paid services to create awareness for the blog, to optimise and embellish display, to track and monitor usage patterns, to manage traffic and even to monetise the blog via participation in the FeedBurner Ad Network.

38 For a full explanation of Google's business model see https://www.google.com/adsense/afc-online-overview or http://en.wikipedia.org/wiki/AdSense.

39 In Second Life users and firms can advertise through buying land and building firm-specific buildings, locations or shops. In theory they can also create standard advertisements on their land.

40 By posting the content the sites receive a limited irrevocable, perpetual, non-exclusive, transferable, fully paid-up, worldwide licence (with the right to sublicense) to use, modify, publicly perform, publicly display, reproduce, and distribute such content through the particular site.

41 See http://us.cyworld.com/mall/index.php for the English language version CyWorld mall and http://cyworld.nate.com/mall/mall5_index.asp for the Korean gift shop.

42 Excluding impacts on businesses from the internal or external use of participative web technologies themselves (*e.g.* by disseminating business news internally via a blog).

43 Data obtained from the Consumer Electronics Association, 'Consumer electronics growth to continue through 2007 according to new CEA Forecast 2005", press release, 14 August 2006, and CEA study on US household ownership of Consumer electronics products. The Consumer Electronics Association forecast total factory-to-dealer sales in the US to reach USD 140 billion in 2006, 8% growth over 2005. See http://consumerelectronicsdaily.typepad.com/consumer_electronics_dail/2006/08/ce_mar ket_to_gr.html. See also OECD (2005b) for the impacts of digital content on the consumer electronics industry.

44 'YouTube success to spawn US$2b video ASP boom", in: *IT News,* 17 October 2006, available at: www.itnews.com.au/newsstory.aspx?CIaNID=40872.

45 'Your Tube, Whose Dime?" in *Forbes,* 28 April 2006, available at: www.forbes.com/intelligentinfrastructure/2006/04/27/video-youtube myspace_cx_df_0428video.html.

46 See Limelight financial investors release at www.limelightnetworks.com/news/pr.2006.07.20.html.

47 In May 2005, Japan had 115 providers (including SMEs) providing blog services and 75 providers of SNS. See www.soumu.go.jp/joho_tsusin/eng/Releases/Telecommunications/pdf/news050517_2_1. pdf.

48 Other findings of the 'Generator Motivations Study' of the Interpublic Emerging Media Lab at http://ipglab.com/ include that as many as 73% of content generators notice Internet advertising, a much higher share than the male 18-24 year-old group as a whole. Also, 57% of all content creators surveyed said they are willing to feature brands in their videos, and many within the group have already done so. 62% are motivated by personal recognition, while 60% are driven by cash compensation.

49 See http://hcs.harvard.edu/cyberlaw/syllabus/dmx.pdf for more explanations.

50 This system faces a number of problems including leakage of music from the DMX onto traditional P2P networks, a form of click fraud as artists have other sites downloading their content to generate revenue for these other sites rather than for DMX and its artists.

51 This model requires that a user must be online while viewing the content to generate advertisement revenue. It does not work well for videos that are downloaded and then viewed later offline or on a portable audio/video player.

52 The *International Herald Tribune*, for instance, signed a deal in May 2006 to syndicate content from the Korean citizen journalism site OhmyNews.

53 'Korea: SK Com. to launch new search engine", in *Korea Times,* 23 October 2006.

54 See http://dovecreamoil.com/.

55 Budweiser, for instance, allows users to put words in the mouth of celebrities or animals to send as postcards. See http://veepers.budweiser.com/service/Start.

56 Examples are the folk singer Sandi Thom and the Artic Monkeys.

57 Coca-Cola and iTunes have recently announced a web page to foster new musical talent (artists can send their music to a selection committee for airing).

58 See www.desiresdavenir.org/. This was an important part of candidate Ségolène Royal's campaign.

59 See http://blog-ump.typepad.fr/.

60 See www.loiclemeur.com/english/2005/12/nicolas_sarkozy.html.

61 Collaborative educational initiatives such as MIT OpenCourseWare provide opportunities for educators and students to make use of these resources and improve education. The University Channel podcast (Princeton University) makes videos of academic lectures available.

62 On the changing role played by broadcasters in shaping politics and the democratic process, see the ongoing conferences and seminars 'Beyond Broadcast 2006: Reinventing Public Media in a Participatory Culture', Harvard Law School, 24 February, available at: http://cms.mit.edu/.

63 This section benefited from research conducted by Elizabeth Stark, consultant, United States.

64 The analysis is based on the terms of service and the Privacy Policy of a sample of 15 widely internationally used UCC sites: Flickr, Ofoto, BoingBoing, Digg, del.icio.us, Bebo, Cyworld, Facebook, Friendster, MySpace, Orkut, Dailymotion, Google Video, YouTube and Second Life. The tables are a generalisation of their terms of service. Other UCC sites, especially non-English-speaking ones, may have different terms of service. Terms of service vary according to local legal frameworks and cultures.

65 The final provision of the *OECD Council Recommendation on Broadband Development* is the '[e]ncouragement of research and development in the field of ICT for the development of broadband and enhancement of its economic, social and cultural effectiveness".

66 See also the OECD-BIAC Workshop on Future of Online Audiovisual Services, Film and Video: Issues for Achieving Growth and Policy Objectives, 29 September 2006, Summary, DSTI/ICCP/CISP/IE(2006)2, available at: www.oecd.org/dataoecd/33/42/37866987.pdf.

67 See also comments and suggestions by the UK regulator OFCOM supporting audiences being able to re-use content, available at: http://www.ofcom.org.uk/media/news/2007/01/nr_20070124a.

68 See also the *OECD Council Recommendation on Broadband Development,* available at: www.oecd.org/dataoecd/31/38/29892925.pdf.

69 Regulators will have to monitor whether their current approaches to guarantee competition in the telecommunication market will be adequate for new NGN environments. See the OECD NGN Forum available at: www.oecd.org/document/12/0,2340,en_2649_34223_37392780_1_1_1_1,00.html.

70 This technological set-up is based on historical Internet use patterns which did not originally show large user demand for high upload capacity.

71 See the *OECD Council Recommendation on Broadband Development,* available at: www.oecd.org/dataoecd/31/38/29892925.pdf.

72 See also the OECD-BIAC workshop, in footnote 66.

73 Comments from Prof. Dr. Urs Gasser, Dr. Martin Senftleben and OECD Delegations on this section are gratefully acknowledged.

74 The requirement of originality can be understood as a reference to the uniqueness of works resulting from the personal and individual character of the process of creation (WIPO, 2006a).

75 WIPO Copyright Treaty (WCT) and the WIPO Performances and Phonograms Treaty (WPPT) at www.wipo.int/copyright/en/treaties.htm.

76 OECD (2004), *Recommendation of the Council on Broadband Development*, C(2003)259/FINAL, available at: www.oecd.org/dataoecd/31/38/29892925.pdf.

77 See http://fairuse.stanford.edu/Copyright_and_Fair_Use_Overview/chapter9/index.html.

78 See also Hugentholtz (1997) for a discussion of fair use in different legal regimes.

79 WIPO Copyright Treaty (Article 10) and WIPO Performances and Phonograms Treaty (Article 16), and Agreement on Trade-Related Aspects of Intellectual Property Rights TRIPS (Article 13).

80 Agreed Statements concerning Article 10 of the WIPO Copyright Treaty adopted by the Diplomatic Conference on December 20 1996.

81 17 U.S.C. 107, U.S. Copyright Act (found in Title 17 of the United States Code).

82 Australian Copyright Act, Division 3, Art. 40 ff.

83 UK Copyright, Designs and Patents Act 1988, Chapter III.

84 See EUCD Article 5, para. 3 and 4. Exceptions are valid for *(i)* teaching or scientific research, *(ii)* for the benefit of people with a disability, *(iii)* reproduction by the press, communication to the public or making available of published articles/broadcasts on current economic, political or religious topics or of broadcast works or other subject-matter of the same character, *(iv)* quotations for purposes such as criticism or review, *(v)* use for the purposes of public security or to ensure the proper performance or reporting

of administrative, parliamentary or judicial proceedings, *(vi)* use of political speeches as well as extracts of public lectures or similar works or subject-matter to the extent justified by the informatory purpose and provided that the source, including the author's name, is indicated, except where this turns out to be impossible, *(vii)* use during religious celebrations or official celebrations organised by a public authority; *(viii)* use of works, such as works of architecture or sculpture, made to be located permanently in public places; *(ix)* incidental inclusion of a work or other subject-matter in other material, *(x)* use for the purpose of advertising the public exhibition or sale of artistic works, to the extent necessary to promote the event, excluding any other commercial use, *(xi)* use for the purpose of caricature, parody or pastiche, *(xii)* use in connection with the demonstration or repair of equipment; *(xiii)* use of an artistic work in the form of a building or a drawing or plan of a building for the purposes of reconstructing the building *(xiv)* use by communication or making available, for the purpose of research or private study, *(xv)* use in certain other cases of minor importance where exceptions or limitations already exist under national law.

85 EUCD, Article 5. See Senftleben (2004), for more on this topic.

86 See Japanese Copyright Act, Subsection 5, Korea Copyright Act, Section 6.

87 Korea Copyright Act, Section 6, Article 26 (Public Performance and Broadcasting for Non-Profit Purposes): (1) It shall be permissible to perform publicly or broadcast a work already made public for non-profit purposes and without receiving any benefit in return from audience, spectators or third persons: Provided that this shall not apply to cases where the stage performers are paid any normal remunerations. (2) It shall be permissible to reproduce and play for the general public any commercial phonograms or cinematographic works, if no benefit in return for the relevant public performance is received from audience or spectators: Provided that this shall not apply to the case as prescribed by the Presidential Decree. Japanese Copyright Act, Subsection 5, Japan: Article 38 (Public Performance and Broadcasting for Non-Profit Purposes. *i.e.* works can be performed or exhibited freely if the performer is not remunerated, and the audience is not charged an admission (1) It shall be permissible to publicly perform, present and/or recite a work already made public, for non-profit-making purposes and if no fees are charged to the audience or spectators ('fees' includes consideration of any kind whatsoever for the offering and the making available of a work to the public; the same shall apply below in this Article), to audiences or spectators. The foregoing, however, shall not apply when the performers or reciters concerned are paid any remuneration for such performance, presentation or recitation.

88 Inaugural Internet Governance Forum Meeting Athens, Greece, 30 October-2 November 2006, Content Rights for the Internet Environment, available at: www.intgovforum.org/Athens_workshops/Content%20Rights%20workshop%20report.pdf.

89 Viviane Reding, Commissioner for Information Society and Media, "The Disruptive Force of Web 2.0: how the new generation will define the future", Youth Forum, ITU Telecom World, Hong Kong, China, 3 December 2006, available at: http://europa.eu/rapid/pressReleasesAction.do?reference=SPEECH/06/773&format=HTML&aged=0&language=EN&guiLanguage=en.

90 See UK Treasury (2006) and the discussion on orphan works in many OECD countries.

91 Recommendations 11 and 12.

92 For example, bloggers who posted audio clips of part of a copyrighted broadcast to criticise the content and tone of the programme were asked to take down relevant material. See www.chillingeffects.org for an attempt to take stock of such notices.

93 See also Regner *et al.*, (2006) on flexible platforms for free content created by users.

94 By posting the content the sites receives a limited irrevocable, perpetual, non-exclusive, transferable, fully paid-up, worldwide licence (with the right to sublicense) to use, modify, publicly perform, publicly display, reproduce, and distribute such content through the particular site.

95 Universal Music filed a lawsuit against Grouper and Bold (two video sharing sites) for violation of their copyright arguing that the latter services host their content without authorisation. In some cases, such lawsuits or related concerns have led to video-sharing sites removing videos from the sharing site.

96 For an example of such take down notice procedures for a user-created content site see: http://secondlife.com/corporate/dmca.php.

97 Section 512(c) of the US Digital Millennium Copyright Act.

98 Section 512(a) of the US Digital Millennium Copyright Act

99 Section 39B of the Australian Copyright Bill.

100 Directive 2000/31/EC of the European Parliament and of the Council of 8 June 2000 on certain legal aspects of information society services, in particular electronic commerce, in the Internal Market ('Directive on electronic commerce'), Article 12. It is implied that the online intermediary's 'activity is of a mere technical, automatic and passive nature, which implies that the information society service provider has neither knowledge of nor control over the information which is transmitted or stored' (Recital 42).

101 'Directive on electronic commerce', Art. 15.

102 'Directive on electronic commerce', Recitals 40 and 46.

103 'Directive on electronic commerce', Art. 16.

104 In March 2005, the Italian government promoted the San Remo charter for the adoption of a co-ordinated set of codes of conduct by the content industry, ISPs, network operators, manufacturers and rights owners, to foster the availability of quality content in a secure environment, and to organise and promote educational campaigns in particular amongst youth to ensure the respect of digital rights. See for France, the *Charte d'engagements pour le développement de l'offre légale de musique en ligne, le respect de la propriété intellectuelle et la lutte contre la piraterie numérique*, available at: www.culture.gouv.fr/culture/actualites/conferen/donnedieu/charte280704.htm. At EU level see the European Charter for Film Online, available at: http://ec.europa.eu/comm/avpolicy/docs/other_actions/film_online_en.pdf.

[105] United States Supreme Court Decision in MGM Studios, Inc. v. Grokster, Ltd. 545 U.S. 913 (2005), at: http://caselaw.lp.findlaw.com/cgi-bin/getcase.pl?court=US&navby=case&vol=000&invol=04-480.

[106] See WIPO (2005).

[107] See http://blogs.law.harvard.edu/palfrey/2007/02/02/viacoms-cease-and-desist-letters-for-a-home-video/.

[108] Recommendations 15 and 16.

[109] The Geneva Declaration of Principles and the Tunis Commitments of the World Summit on the Information Society refer to and underline the importance of freedom of expression and the free flow of information, ideas and knowledge and the identification of the appropriate enabling legal, policy and regulatory frameworks that preserve openness as one of the key founding principles of the Internet.

[110] 'Microsoft survey: Blogging Phenomenon Sweeps Asia, According to New Research from Windows Live Spaces', in: *Xinhua-PRNewswire,* 28 November 2006, based on Microsoft surveys of its MSN and Windows Live Online Services Business.

[111] For example, this is also the form chosen by Wikipedia.

[112] Debated in the context of the revision of the UK Violent Crime Reduction Bill. See "Move to ban happy-slapping on the web", *The Guardian,* 21 October 2006.

[113] Certain user-created content services have also implemented special zones for underage users (*e.g.* Teen grid for 13-17 year olds in Second Life).

[114] Even this is not perfect, though, as one could obtain access to such an address and start a profile.

[115] In Korea the posting of pictures and videos of the behaviour of some people has led to 'public shaming'.

[116] See www.oecd.org/sti/security-privacy for OECD work on privacy and ICT security.

[117] Since the early 1990s, several clinics have been established in the United States to treat heavy Internet users. They include the Center for Internet Addiction Recovery, in Bradford, Pa., and the Connecticut-based Center for Internet Behavior.

[118] 'Developing teaching curriculum to disseminate a sound Internet culture", *Korea Herald* (15 September 2006), www.asiamedia.ucla.edu/article-eastasia.asp?parentid=5298.

[119] Survey of 3 000 Internet users. See: www.itu.int/wsis/stocktaking/scripts/documents.asp?project=1120746255&lang=en.

[120] See Republic of Korea, Ministry of Information and Communication, The Center for Internet Addiction Prevention and Counselling, available at: www.itu.int/wsis/stocktaking/scripts/documents.asp?project=1120746255&lang=en.

[121] Statistics Canada (2006) notes that, '[t]hus, it is not that people are becoming anti-social; it is that people are becoming differently social.'

[122] 'Splogs', a combination of blog and spam, are weblog sites with faked articles which the author uses only for promoting affiliated websites or other content, such as stocks.

[123] According to Technorati, approximately 8% of all new blogs are spam, with short-term spikes up to 30%.

[124] See www.oecd.org/sti/spam for the work of the OECD on spam and the OECD Report of the OECD Task Force on Spam: Anti-spam Toolkit of recommended policies and measures, available at: www.oecd.org/dataoecd/63/28/36494147.pdf.

[125] This tag, which can be used in blog comments, would not raise a particular site's ranking within a search engine when it is linked to by spammers.

[126] The authors thank David Holmes (OECD Centre for Tax Policy and Administration) for useful discussions on this topic. See also Joint Economic Committee of the US Congress: www.businessweek.com/innovate/content/may2006/id20060502_832540.htm and Washington Post at http://tinyurl.com/ve34c.

[127] For OECD work on taxation and e-commerce see: www.oecd.org/topic/0,2686,en_2649_33741_1_1_1_1_37427,00.html.

[128] Japan presented the results of its Annual Survey of Digital Content at the December 2006 session of the OECD Working Party on the Information Economy and called for international collaboration in this area.

OECD PUBLICATIONS, 2, rue André-Pascal, 75775 PARIS CEDEX 16
PRINTED IN FRANCE
(93 2007 03 1 P) ISBN 978-92-64-03746-5 – No. 55805 2007